As-salted
A Story of Salt

"The recipe for perpetual ignorance is: Be satisfied with your opinions and content with your knowledge." –Elbert Hubbard

"There is only one good knowledge and one evil ignorance."– Socrates

"He who joyfully marches to the music in rank and file has already earned my contempt.
He has been given a large brain by mistake, since for him the spinal cord would surely suffice." –Einstein

Table of Content

Introduction

Chapter One—Our Tragic Health

Chapter Two—Misconceptions

Chapter Three— War on Salt

Chapter Four— Goodness of Salt

Chapter Five— Chemistry of Salt

Chapter Six—Low Salt Symptoms

Introduction

Salted...... Deceived and Unaware

In our busy lives we rush about doing the best we can, but sometimes we fail to stop and take a hard look at what we're actually doing. Without discernment of some of the things we are doing, we lack confidence to make wise healthy choices. We suffer but keep moving forward, never asking why.

Our health of millions in America is declining. Some health problems are altogether avoidable; many others can be vastly improved; some are the result of being misled, while many people have misconceptions about what is healthy. There is a mysterious fact that individuals feel an innate ability to "know" what is healthy without studying the details.

Take a simple thing like salt. Why would we stop to analyze such a basic substance? Why would we doubt anything we are told in magazines, doctors' offices and by the FDA? Maybe we *should* scrutinize it. What we do not understand about salt could ruin our health and your life.

There are many misconceptions we have been unaware of with big names like high fructose, and hydrogenated oil and carbohydrates have been brought to our attention. Lately, even things like genetically engineered food (GMOs) are beginning to be revealed for their inherent dangers.

Salt has a story. We have been misled. We do not understand it. It is not what we think. Logic does not work when we are not told the truth. You need the truth.

Salted An as-salt on our common sense...

Chapter One—Our Tragic Health

How have we been as-salted?

Unless you have been on the moon for the last few years, you are aware that our health care system is in decline. Oh, there are a lot of great surgeries and new discoveries. I personally love those tests that tell you everything, if you can afford them. Our longevity charts soar to new heights at 78. Hey, I'm going to 98 years old, aren't you? Many honest humans have tried hard to be healthy, pushing the envelope to giddy accomplishments like "50 is the new 30" or is it "60 is the new 30"?

Take a step back and look again. Oh, come on. Step back enough to open your mind and see the view as an intelligent bystander. You and I see cancer still running rampant, good people and loved ones suffer from all kinds of ailments.

- ADHD has increased 500% since 1991[i]
- Arthritis is up 2.5% -- per year[ii]
- Thyroid disease has doubled[iii]
- ALS has increased in deaths 400%[iv]
- Asthma is an epidemic, rising by 42% in ten years and killing 5,000 per year[v]
- Autism is up 18% per year[vi]
- Depression is up 63%[vii]

- Suicide is up 109%[viii]
- Diabetes has increased tenfold in the last five years![ix]
- High blood pressure has increased by 30%[x]
- Obesity is up another 35% and predicted soon to be 50% of Americans----seen any of these guys? Hey still look it up if you haven't noticed (footnote)[xi]
- Celiac disease has gone up four fold in fifty years and sits at 1% of the population.[xii]

I am sure you have noticed how long the lines are at the pharmacy. One half of all Americans are taking prescription drugs, a ten per cent increase, while those taking more than five prescriptions have risen more than seventy per cent. [xiii]What happened? What have we done? Where have we gone wrong? The list could go on and on; however, we must move on.

SALT: FACTS VS. MISCONCEPTIONS

Scientists and media are telling us to eat less salt and that it is bad for us. This book will explain how that assumption is wrong. There is so much misunderstanding about this vital substance.

The science of salt has been vague and misleading. We will clarify the many misconceptions about salt and give details on how to live a healthier existence. Here is a list of some of the facts and questions you will find out about salt as you read through this book:

1. The clarification of what salt is and what it does.
2. The miracle of salt.

3. How salt is actually made. Different from what you think.
4. How salt interacts in your body and how it affects you.
5. What kind of salt should I use?
6. Explanation of "salt sensitivity". What is "salt sensitivity"? Should I care?
7. What health benefits should I expect using the good "natural salt"?
8. Science of salt and how bad salt and good salt differ.
9. Does "sea salt" still have the same side effects as the table salt?
10. How about "salt substitutes"? Can I just use them instead? Should I care?
11. Doesn't salt cause water retention and swelling?
12. Does salt cause high blood pressure and heart disease?
13. How does salt work biologically in the body?
14. How scientists have performed testing for effects of salt.
15. How much salt is safe? Do we even need it?
16. How about "salt substitutes"? Can I just use them instead?
17. Doesn't salt cause water retention and swelling?
18. Does salt cause high blood pressure and heart disease?
19. Haven't scientists performed testing for effects of salt?
20. What kind of salt should I use?
21. How much salt is safe? Do we even need it?
22. What is salt? What makes it? Will there be a shortage of salt some day?

I look forward to traveling through this short book on salt, however to establish a full understanding and a proper perspective, we should first look at other similar things that have been misunderstood. Our everyday belief system is built on what we have been told, assumed or interpreted from information. Often we have come up with the wrong answer.

*P*lease read through the following chapter with examples of how we might have misconceptions in our everyday perceptions. Feel free to disagree. I only ask you to review them. I have spent many hours researching and have tried to be as precise and concise as possible. The subject of SALT continues with Chapter Three.

Quick Fact: *I will interject a quick fact before going on to the next chapters. Where does all salt come from?   THE SEA!   (Details in Chapter Fifteen)*

Chapter Two – Misconceptions

The pernicious consequence of misinformation being rammed at us regarding health is devastating. This chapter will travel through the many general misconceptions we experience in our daily lives to show the mind set of how we also have misconceptions about our health. At times, our instincts kick in and we get it right; however, other times, we find that we are so monumentally wrong that it requires some mental adjustments. It just might take a little time for more details and diverse information to sink in.

Sometimes, even when we *thought or believed* information we understood was true, we later found the truth is different than anything we could have imagined. Whether we did not have the full information, the latest knowledge, had misinformation or were purposely deceived, it is time to open our minds to some possible errors of judgment and consider the world around us. It is time to open our minds and set the record straight about salt.

Since there is a huge misconception regarding salt, our purpose here is to run through some misperceptions and illustrate how easy it is to fall for the delusion when the truth was something else. I hope you take a moment to read through this chapter just to open our minds to reality vs. fallacy.

First let me "vent" my frustration of missed opportunity for mankind. We are blessed.

We are currently experiencing a "Goldilocks" era on this earth. Forget the fact the earth is cruising the galaxy at just the right distance from its core to keep us from being zapped with too much energy by gamma rays or deprived of too little energy by being too far out. Also, let's overlook the fact the Earth is about the right distance from the sun and we have an electromagnetic field protecting us from the harsh solar wind that would render us into another desolate planet like Mars.

Some scientists say we have been living in an Earth "summer" since the Holocene period 12,000 years in the past and even more so since 5800 B.C. Whether this is the exact time frame, that is when the Atlantic circulation turned on and the westerly's resumed. We are now in an interglacial warm age. Throughout the ages, the earth's "wobble," its position in the Galactic plane and the solar activity of the sun have caused many cycles on earth.

A slight change in the sun's activities and we enter a "Dark Age" as we did in 1200 B.C. and 500 B.C. In Rome in 1,000 A.D. lost 97% of its inhabitants due to a mini cooling off period. During these times crops were poor or non-existent, sending the populists' health plummeting. The decline in population, civilization, financial stability, and even religious organizations followed mini earth winters.

Best of Times

We take for granted the "best of times" conditions we are experiencing on earth today with its long growing seasons, mild/warm weather, healthy conditions and prosperity to last forever.

However, as Jupiter and Saturn now begin to tug the Earth into the next elongated orbit away from the sun and the moons gravity starts the Earth to tilt again away from the sun, mans' existence will begin to be challenged. Nothing can stop this planetary motion. No amount of global warming will stave off the eventual return of colder weather.

The sunlight will be reduced in its intensity and greatly diminished from the most northern places such as Alaska, Canada, Siberia and of course, Greenland and Iceland where a new year-round snow cover will appear.

Cooler shorter summers will cause short growing seasons, pollination to be difficult and crop yields to become meager. Millions of people will suffer or die from famines as in France in the seventeenth century. With poor nutrition also comes decreased health, leaving man unable to survive diseases. Even the fish of the sea will struggle with the cooling affects, just as the trees, the plants and the animals.

Not only will economic strife be a certainty as in the past cooling period of 1560-1850, social upheavals will appear as multitudes are starving.

So, the point is, we are presently in the best of times on planet Earth and yet we cannot get it right! If we cannot "get it" correct now, then when? We have enjoyed the best that this solar system and galaxy has to offer and we have thrown away our greatest chance to perfect our health. I have been under the delusion everyone was trying hard to make things healthy and better. We have settled for accepting the misconceptions, lies and lack of common wisdom. Complacency, peer pressure and lack of knowledge have ruined our health. Greedy industries and politicians have poisoned the food! Where did we get off the wagon of common sense?

I have had my own misconceptions that have rocked my mind and as disconcerting as it is, I have been misled by the fundamental pillars of our *advanced* society.

Misconceptions recognized as wrong: Use *endnotes for* further information. (*xvi, xv, xviii etc)*

- Margarine is better for you than butter (margarine is hydrogenated, includes free radicals, synthetic vitamins, emulsifiers and preservatives, hexane and other solvents, bleach, artificial flavors, and sterols)[xiv] [xv]
- Vegetable oil is good [xvi](not even canola. Except olive oil, oils are hydrogenated and chemically extracted)
- Eggs are bad for you[xvii] (eggs are a great source of protein and nutrients)
- Blueberry muffins are healthy to eat (sugar, wheat)
- Sports drinks are better than water[xviii] (sugar, not enough electrolytes)
- Sugar free[xix] is healthy with its aspartame, saccharin, high fructose, and Splenda (weight gain, addictive, aggravates diabetes, contains methanol, they are carcinogens [xx] and damages DNA[xxi])[xxii] [xxiii] [xxiv]
- Low-carb diets like Adkins diet works[xxv] (fools body temporarily)
- Fat free or low fat[xxvi] is better than EFA[xxvii] like olive oil (low fats are unhealthy trans fats and hydrogenated while

Essential Fatty Acids (EFA) are essential. EFA like fish oil or olive oil has many health benefits like lowering blood pressure) [xxviii]

- Hydrogenated [xxix] and polyunsaturated fats [xxx] [never consume hydrogenated] [avoid unsaturated]
- Sugar was harmless (Hmm. However I still love it)
- Tanning booths are healthy [xxxi] (The UV rays damages the skin, increases odds of cancer 2.5X and detrimental to the immune system. Check out the web site shown)
- The government can spend our money better than we can [congress costs $6B/yr[xxxii] for salaries and perks] [government spends $2.5T/yr][xxxiii]
- We are alone in the universe *(can't prove this one)*
- Frying meat/food is okay (creates carcinogens and contributes to cancer) [never deep fry[xxxiv]] [do not eat burnt food[xxxv], [xxxvi]]

More Misconceptions... (w/ clarification):

Climate

Whether the earth is warming or changing and no matter where you are on this subject, please take a moment to consider some of the science and consider the many things accounting for changes in our weather other than carbon emission. Invest in wind mills and solar panels? Sure, let's do *some* solar and wind. It might clean the air up a little. We all need cleaner air.

But borrow trillions of dollars from China to buy the windmills from overseas and buy solar panels from China? Go bankrupt paying for a dream we do not have enough information on yet? Will this change global warming? Not likely. It is not produced, maintained and operated without expending other energy sources and it is definitely not free. They also have their own environmental affect. Windmills kill 573,000 birds a year in the US.[xxxvii]

Trees are carbon

The Earth has a marvelous system intact to protect itself and cleanse itself. Take for example trees. A tree is a very large chunk of material. The tons of material that makes up a large tree is not from the ground. There is no big sink hole around each tree where it has sucked up the surrounding earth and used it for building itself toward the sky. Trees are made of................carbon. The same carbon dioxide you breathe out. Trees and forests suck up huge amounts of CO_2 and spew out oxygen. [xxxviii]
Please read...*http://www.npr.org/blogs/krulwich/2012/09/25/161753383/trees-come-from-out-of-the-air-says-nobel-laureate-richard-feynman-really*

Wouldn't it be good to invest money into saving forests and replanting around the world? Helping the earth to help itself is a very efficient plan.

Green

Some "green" ideas and efforts are noble, like limiting plastic, things that go to the landfill, burning trash, limit our wasting and limit polluting in all forms. However, each and every action needs scrutinized and studied. Is recycling "green"? Some things are worth recycling but some are not. Some recycling requires more energy and waste to do so. You must choose on each item yourself and decide. Here are two sites to begin your investigation:[xxxix], [xl]

Politics affect decisions

"Going Green" might be a lot about politics. According to a report (http://abcnews.go.com) by ABC World News, very little of the money appropriated for solar and wind power was spent here in the states! Nearly $2 billion in money from the American Recovery and Reinvestment Act has been spent on wind power last year. The disconcerting fact is the study found that 80 percent of that money has gone to foreign manufacturers of wind turbines.[xli] For now let's forget the fact Europe and others in the world have found wind powered electric generators do not work in the long run; but rather, consider, how much knowledge we actually have about the climate from the less than one hundred years we have been studying it. [Twenty years with satellites]

Is the earth warming? Looks like the arctic ice has been melting and the weather sure is strange. However, there are numerous scientific studies by a lot of respectable scientists that say, "Not so fast". Their facts show something else. Professor Phil Jones from the University of East Anglia's Climatic Research Unit states that we are in fact not experiencing a global warming.[xlii] He also points out that no grant money has been awarded to look for the *natural* greenhouse effects. This seems a little lop-sided.[xliii]

Bestselling author and former NASA climatologist Roy Spencer, in his book "The Great Global Blunder," reveals how climate researchers have mistaken cause and effect and been fooled into believing the earth is more sensitive to human action and carbon dioxide than it truly is.[xliv]

The earth has not warmed in two decades. We have no historical temp data for 70% of the earth's surface (oceans) and 99% of the earth's existence! Scientists cannot model a planet's climate with less than 1% of the data.

Today satellites only capture radiant temps which are meaningless for determining global temperatures. The worldwide buoys with temp probes placed one meter below the surface have only been deployed less than five years. We need to be very cautious with what we do with this information.

BTW: In the Sahara desert, the nomads claim there has never been as much rainfall as in the past few years and they have never seen so much grazing land.

> "Now you have people grazing their camels in areas which may not have been used for hundreds or even thousands of years. You see birds, ostriches, gazelles coming back, even all sorts of amphibians coming back,"[xlv] [Plenty of time to buy. (Still 3.6M square miles of sand)

You might want to ask some people around the world if they are experiencing warmer weather.
- 2012 was the coldest winter in over 70 years in Russia.[xlvi]
- 2012 Italy had the lowest temperatures in 27 years[xlvii]
- 2012-2013, 2013-2014, 2014-2015 is predicted to have significantly lower temperatures for the Midwest United States according to AccuWeather.com [xlviii]

Well, of course one little solar cycle can change things, a solar flare can warm us up and a sudden volcano can send us into another cold "mini ice age".
- The Earth is about to enter a "Little Ice Age" according to geophysics Victor Herrera of the University of Mexico.
- We could be in for a cooling period for 200 years, according to the Pulkovo Observatory in Russia.
  It might be more likely we are heading toward a global cooling period as early as 2030. [xlix] But first, let's discuss the *big bad guy*, ...carbon or CO_2.

CO_2 (Carbon Dioxide)
Carbon levels have little effect on climate change!

All government agencies including; IPCC (Intergovernmental Panel on Climate Change), NOAA and NASA, and all funded climate scientists currently monitor the "*greenhouse gases*" by keeping track of methane, nitrous oxide, HCC's (hydrocarbon condensate) and carbon dioxide. The first three air born substances have changed very little since 1990.[l] *(Endnote gives site with charts)* Most of the change has been in carbon levels from approximately 22,000 MMT (million metric tons) to 28,000 MMT.[li]

Stop the clock, shut down the plants, park the cars and bury the farm animals! *We have carbon!* Very scary, right? Take a breath. Beside the fact that the trees love the carbon, it does little to add to global warming, cooling or change. We will now pursue the science of carbon and why it has little to do with global warming.

CO₂ measurements are inaccurate!

Any reference to C02 for dates before 1950 is bogus. Ice core samples are being used by scientists to prove CO2 levels have risen sharply. Ice core samples can be useful for different scientific information; however, for determining levels of carbon in the air, they are just not accurate.

The dubious samples of gas measurements brought up from ice cores show a rather steady increase in CO2 levels, conveniently corresponding to the preconceived idea that increasing industrial activity has produced a steady CO2 increase. As Dr. Zbigniew Jaworowski, former senior adviser to the Polish radiation monitoring service, has shown, the gaseous inclusions in ice cores have no validity as historical proxies for atmospheric concentrations. The continual freezing, refreezing, and pressurization of ice columns drastically alters the original atmospheric concentrations of the gas bubbles. [lii]

CO_2 accounts for only 2% of greenhouse effect

According to the greenhouse warming theory, the increase of atmospheric CO_2 concentration caused by human activity, such as burning of fossil fuels, acts like the glass in a greenhouse to prevent the re-radiation of solar heat from near the Earth's surface. Although such an effect exists, *carbon dioxide is low on the list* of greenhouse gases, accounting for at most 2 or 3 percent of the greenhouse effect. By far the most important greenhouse gas is **water vapor.**[liii],[liv]

In fact, Dr Spencer Weart (PhDs in physics and astrophysics, American Institute of Physics (AIP) in College Park, Maryland, USA) along with other scientists states that the rapid variations seen in the ice cores had been misinterpreted. They did not reflect changes in atmospheric CO_2, but only changes in the ice's acidity due to dust layers. Something had indeed changed swiftly — not the CO_2 level, but the *dustiness* of the entire Northern Hemisphere, as a change in weather patterns swept more minerals from deserts.[lv]

Recent carbon increase report

In a May, 2013 report from the observatory in Mauna Loa, Hawaii, it was stated CO_2 reached 400ppm for the first time since the recording began in 1950. Articles are popping out about comparing the current observations to records from thousands and millions of years ago using the gas bubbles in the ice cores in the arctic. When someone says…"higher than 3 million years" or "higher than 800,000 years" understand that this report you are reading has an *agenda* and is not just reporting facts. These ice core samples are not a valid record for CO_2 and we cannot use them to make huge leaping predictions. Their main message; *let's spend money on things!*

Carbon dioxide is a greenhouse gas. However, I repeat, CO_2 provides about 2% of the Earth's greenhouse effect. Where is this carbon coming from?

Sources of carbon dioxide[lvi]

1. Volcanic activity both under the oceans and on land sends .4 gigatonnes of CO_2 into the air each year. They also release more water vapor than this, which is a much more powerful greenhouse gas (GHG) than carbon.
2. *Farm animals* are responsible for 18% of all GHG emissions! According to the Food and Agriculture Organization of the United Nations (FAO), the animal agriculture sector is a huge contributor at eighteen percent of all the greenhouse gases. (measured in carbon-dioxide equivalent amount as the reference) These numbers do not reflect our pets and other animals on the earth.[lvii] This is more than all combined sources of transportation including cars, trucks, trains, planes and boats!
3. In fact, the farm animal sector annually accounts for 9% of human-induced emissions of carbon dioxide (CO_2), 37% of emissions of methane (CH_4), which has more than 20 times the global warming potential (GWP) of CO_2, and 65% of emissions of nitrous oxide (N_2O), which has nearly 300 times the GWP of CO_2.[lviii]
4. Wildfires. Forest fires in Indonesia in 1997 released up to 2.57 gigatonnes of CO_2 into the atmosphere. This is 40% of the annual emissions from burning fossil fuels!
5. US Forest fires. Thanks to some reckless persons, 80% of all wildfires in the US are caused by man with a whopping 25% of forest fires being started deliberately!
   a. According to National Science Foundation, Christine Wiedinmyer of NCAR and Jason Neff of the University of Colorado, used satellite

observations of fires to estimates carbon dioxide emissions based on the mass of vegetation burned.

b. Their study estimated that U.S. fires release about 290 million metric tons of carbon dioxide a year, the equivalent of 4 to 6 percent of the nation's carbon dioxide emissions from fossil fuel burning.[lix]

c. "A striking implication of very large wildfires is that a severe fire season lasting only one or two months can release as much carbon as the annual emissions from the entire transportation or energy sector of an individual state," the paper states.

6. Living animals and all cellular activity. (Too involved to process a real number, this is a book about salt) I will say that there is a lot of living things from little squirmy to big animals. I would sacrifice the mosquitoes. Just a personal thing.

7. Human activity is charged for 30 megatonnes of CO_2. This includes all human activity such as electric plants, transportation, homes, industry and your *dad-burn breathing air while living.*[lx]

8. Plant decay and fermenting is not included in the CO_2 equation. The natural decay of organic material in forests and grasslands and the action of forest fires results in the emission of about 439 gigatonnes of carbon dioxide every year, while new growth entirely counteracts this effect, absorbing 450 gigatonnes per year.

9. CO_2 emissions resulting from bioenergy production (fuel made from grain) have traditionally been excluded from most emission inventories. (Hmmm?)[lxi]

It's the Moisture

The "climate change" advocates have been unscientifically convinced into thinking carbon emissions could warm up the planet when moisture in the atmosphere causes most of the heat. Many global warming enthusiasts deny moisture is a greenhouse gas, but consider the explanation of clouds and water changing the weather presented in Roy Spencer's "The Great Global Blunder."

Germany professor, Ernst-Georg Beck of the Merian-Schuler, found that the founders of modern greenhouse theory, Guy Stewart Callendar and Charles David Keeling (a special idol of Al Gore's), had completely ignored careful and systematic measurements by some of the most famous names of physical chemistry, including several Nobel prize winners. Measurements by these chemists showed that today's atmospheric CO_2 concentration of about 380 parts per million (ppm) have been exceeded in the past, including a period from 1936 to 1944, when the CO_2 levels varied from 393.0 to 454.7 ppm. [lxii]

Wheat ...*Ah*... the golden grains

Don't eat wheat! Limit your wheat intake! In William Davis's book, "Wheat Belly"[lxiii], he explains that all of our modern grains have been genetically engineered causing devastating health problems. Wheat has turned out to be the worst of all the bio-engineered grains. It is so bad that one slice of bread causes more of a sugar level spike in our blood (called glycemic index) than a candy bar!

Wheat is addictive, the leading cause of celiac disease, and harms the body so much it should not be consumed in its present form. There is a raging epidemic of wheat intolerance in this country.[lxiv] All the while, the FDA continues to espouse the wonders of eating large amounts of "healthy" whole grains (wheat).

Wheat is not just another carbohydrate. It is the ultimate hubris of modern humans that we can change and manipulate the genetic code of another species to suit our needs. Perhaps that will be possible in the distant future. But not today, or anytime real soon, can we know what bio-engineering is doing to our bodies. Genetic modification and hybridization of the plants we call food crops remain crude science, still fraught with inadvertent effects on the plant itself and the animals consuming them. Earth's plants and animals exist in their current form because they are in balance. They have proven their interaction through millions of years.

We step in with the ludicrously brief period of the past half century, alter the course of a plant that thrived alongside humans for millennia, only now to suffer the consequences of our shortsighted manipulations. In the 10,000-year journey from innocent, low-yield, not-so-baking-friendly einkorn grass to high-yield, created-in-a-laboratory, unable-to-survive-in-the-wild, suited-to-modern-tastes dwarf wheat, we've witnessed a human engineered transformation. Perhaps we can recover from this disaster called bio-engineering, but first we need to recognize what we've done to this obsession called "wheat."[lxv]

Three of the processes that go on inside your body that have a *major* impact on your rate of aging are called "glycation", "inflammation", and "oxidation." When we talk about aging, we're not just talking about wrinkles on your skin or how thick your hair is... we're also talking about factors you can't see, such as how well your organs function, and whether your joints are degrading.

This deals with "glycation" in your body, and substances called *Advanced Glycation End Products* (AGEs). These nasty little compounds called AGEs speed up the aging process in your body including damage over time to your organs, your joints, and of course, your skin.

So with that said, what is one of the biggest factors that increases production of AGEs inside your body? This may surprise you, but high blood sugar levels over time dramatically increase age-accelerating AGEs in your body. This is why type 2 diabetics often times appear aged, looking older than their real age. But this age-increasing effect is not just limited to diabetics. It can affect everyone.

That is where "whole wheat" comes in.

Cover-ups by massive marketing campaigns of giant food companies want you to believe that "whole wheat" is healthy for you... but the fact is, wheat contains a very unique type of carbohydrate (not found in other foods) called *Amylopectin-A*, which has been found in some tests to spike your blood sugar higher than even pure table sugar.

In fact, amylopectin-A from wheat raises your blood sugar more than almost any other carbohydrate source on earth based on blood sugar response testing.

This means that wheat-based foods such as breads, bagels, cereals, muffins, and other baked goods often cause much higher blood sugar levels than most other carbohydrate sources. As you know now, the higher your average blood sugar levels are over time, the more AGEs are formed inside your body, which makes you age FASTER.

You've probably heard about the potential health-damaging effects of gluten (mostly found in wheat) in the news recently, but this *blood sugar aspect* is not talked about that often, and is yet another reason to reduce or eliminate wheat-based foods in your diet.

Many people have reported health benefits of reduced or disappeared chronic joint pain from just removing 95% of the wheat from their diets. The added bonus? They also claim they have lost "love handles" and stubborn weight, all from reducing their wheat intake!

Another Problem with wheat based food: As it turns out, baked wheat products contain carcinogenic chemicals called acrylamides that form in the browned portion of breads, cereals, muffins, etc. Yikes-mom was wrong-it does not make you pretty if you eat the crust!

These carcinogenic acrylamides have been linked in studies to possible increased risk of cancer and accelerated aging. Note that acrylamides are also found in high levels in other browned carbohydrate sources such as french fries or any other browned starchy foods.[lxvi]

Marketing or Science?

Sometimes what we believe to be true can be more about advertising and less about science. Multi-million dollar marketing campaigns reach every aspect of our daily lives. Eventually, we just assume that something is true. Loud attractive promotional phrases catch our attention, but many subtle suggestions permeate our subconscious until we just "know" something. No need to look it up, question it or look at the science. We even have the stamped approval of the government telling us what to buy.

Every magazine, talk show, and even doctors repeat what they have heard or been told *without* looking up "peer reviewed science experiment reports!" Often what we "know" is what the lobbyists, drug reps, or manufacturers have insisted is true. Scientific experiments are performed in two ways.

- One is by the manufacturer or subsidized by a biased political system
- The other is by a non-aligned lab that publishes their results to be examined by other scientists and then, after reviewing the results, criticize and leave comments. THIS is what we have referenced in this writing when possible and this is something to believe in. [It is sometimes difficult to read with its science jargon]

We are looking at the very pillars of our society. This time, we are looking right down their throats. We are questioning them, each and every one of them. No longer can we assume pillars like *milk* are healthy just because we have commercials that have told us it was, our moms told us it was or that everybody around us *said* it was.

Milk! Wheat! Sweeteners! We were made to believe they were good.

Corn-based Foods
Corn syrup, corn cereal, corn chips and corn oil have a dark side.
This is quite a variety of stuff you might eat every day... corn chips, corn cereals, corn oil, and also the biggest health-damaging villain that gets most of the media attention, high-fructose corn

syrup (HFCS). Read the ingredients on the box of corn cereal and notice that it is mostly sugar.

Corn-based starchy foods such as corn cereals, corn chips, etc. also have a big impact on blood sugar levels and therefore can accelerate aging.[lxvii]

High Fructose Corn Syrup

But here's another nail in the coffin for corn... it turns out scientists have found the fructose in HFCS causes 10x more formation of AGEs in your body than glucose! Yes, that's right... that means the HFCS you consume daily in sweetened drinks and most other processed foods (yes, even salad dressings and ketchup) contribute to faster aging in your body... as if you needed yet another reason to avoid or minimize HFCS!

We're not done yet with corn... it gets even worse...

Another major issue with corn-based foods and corn oils is these foods contribute to excessive amounts of omega-6 fatty acids to your diet, causing an imbalance in your omega-3 to omega-6 ratio and leading to inflammation and oxidation within your body.

Avoid or reduce corn-based foods like corn chips and corn cereal as much as possible. These aren't as bad as wheat in relation to blood sugar, and they don't contain gut-damaging gluten like wheat does, but they are still bad for you. When it comes to corn syrup or corn oil, avoid as much as you can if you want to stay lean and young looking.

Keep in mind that some starchy foods like white rice, oatmeal, and white potatoes can also have significant impacts on your blood sugar and thereby can increase formation of AGEs in your body. Use starch in smaller portions if you eat them and balance them with healthy fats and protein to slow blood sugar response.[lxviii]

"Vegetable oils"...Soybean oil, Canola oil, or others

These have been marketed over the years by giant food companies as "healthy," but if you understand a little biochemistry about how these highly-processed oils react inside your body, you will quickly see how unhealthy they really are.

First, *anything* labeled soybean oil, canola oil, corn oil, vegetable oil, or cottonseed oil (these are in a LOT of processed foods you probably eat) have undergone a refining process under extremely high heat and use chemical solvents such as *hexane*. (The hexane is used to extract the oils out of the seed)[lxix]

This leaves you with an oil where the polyunsaturated fats have undergone a lot of oxidation and are therefore VERY inflammatory inside your body, producing free radicals, damaging your cell membranes, contributing to faster aging, heart disease, and other possible health problems.

If you want to avoid the health-damaging effects of soybean, canola, corn and other "vegetable oils", instead opt out for **truly healthy oils** and fats such as extra-virgin olive oil, avocado oil, virgin coconut oil and grass-fed butter.[lxx]

Olive oil

Olive oil is made by cold pressing the olive. That is all the processing it takes. Olive oil is much healthier [lxxi] and remains healthier when heated during cooking.[lxxii]

Cow milk

Cow milk does not belong in our body! They feed the animal engineered grain (the cow body was designed to eat grass) that has been overexposed with herbicide, pesticide, and treated with growth hormone and antibiotics. Then they really go to work completely ruining our cow milk. First they pasteurize the milk. Then they homogenize it, radically vibrating it to the point it's unrecognizable on a molecular and atomic level. Since it has become a non food particle, our bodies strain to digest it so much that it must take calcium from our bones to help deal with it. Have you heard of lactose intolerant and have wondered why?[lxxiii] The milk industry pays billions of dollars per year for lobbyists and promotional advertising. Sadly, the FDA also touts this wonder drink.

Please don't take my word for it. Do a Google search for *Hazards of Milk* for a complete explanation. There is a lot of information written on this. [lxxiv]

To add insult to the insult, the vitamin "D" the factory adds to your milk to help make it healthy is not vitamin D. There are two types of vitamin D. The natural D3 contains the same D your body makes when exposed to sunlight. The synthetic is vitamin D2 and is called ergocalciferol.
The book "The Milk Imperative" explains why vitamin D is actually harmful to your health when added to milk.

Note; try Almond Milk. It tastes good and is healthy for you. It is made by squeezing the nuts and adding organic water.

Organic milk

Organic milk is miles and miles better than regular milk. However, if you care enough to make this attempt at health, you should be aware of a few things.

The term "organic" legally means the farm the milk comes from does not use antibiotics to fight infections in cows or hormones to stimulate more milk production. Many organic producers also feed the cows with organic feed without herbicide and pesticide however they still feed them *grain* in lieu of the natural grass they were biologically born to eat.

Organic Milk lasts longer

Organic milk lasts longer because producers heat it more. The process giving organic milk a longer shelf life is called ultrahigh temperature (UHT) processing, heats milk to 280 degrees Fahrenheit for two to four seconds, killing any bacteria in it. This extra sterilizing is necessary due to the further shipping distances from the scarce organic farms. This heat does destroy some of the nutrition of the milk. The heat caramelizes the milk giving it a sweater taste. [lxxv]

Organic Milk scammers

Most organic milk is good and is produced by reputable manufacturers. However, the USDA agency still allows organic milk producers to bring conventional dairy cattle pumped full of hormones and antibiotics onto their farms and call them "organic" cows when they start giving milk.

Some of the big name-brand "organic" milk actually comes from questionable factory farm operations that cut corners and exploit loopholes. The Organic Consumers Association (OCA) has been fighting this for years with public education campaigns, boycotts and legal strategies.

Conventional replacement <u>dairy</u> calves, typically bought at auctions, likely receive antibiotics, toxic insecticides and parasitic-ides as well as conventional feed during their first year of life before being "converted" to organics.

"Real organic farmers don't buy replacement heifers," said Mark A. Kastel, Senior Farm Policy Analyst at The Cornucopia Institute. lxxvilxxvii

Know and trust your supplier.

All descriptions, claims and even lists of ingredients have been scrutinized by the many manufacturers for legal limits and even loopholes. For instance, the term "natural" has no legal or regulatory meaning. That's right, none at all. The FDA has never created a definition for it because "it is too difficult to define due the fact that the food has been processed in some manor and no longer a product of the earth". lxxviii [What?]

Matter of fact, the FDA states *high fructose corn syrup* is "natural". Anyway, the term is meaningless. lxxix What? We all need to do our own research to determine what is and is not natural.

OCA—Organic Consumers Association

Look for the organization OCA to certify it is truly organic products. The label should identify the milk as produced from cows fed with herbicide and pesticide free feed.
Here is a site to look up trusted organic suppliers:
http://www.cornucopia.org/dairysurvey/

USDA/FDA-Working against our Health

Federal Agencies like the FDA and USDA were created to protect consumers; however, they have been so corrupted by big business and bribed politicians they behave much like a branch offices of companies like Monsanto, (which they were charged to regulate).

The officials for the USDA, FDA and EPA are rewarded with lucrative jobs at the companies they were regulating.

Example: Michael Taylor was hired by the FDA as executive assistant to the commissioner after working as the Monsanto lawyer. Then the FDA gave him the sole responsibility of deciding how to label rBGH-derived milk. To keep milk from being stigmatized in the market place he ruled that the labels cannot state a difference between the rBGH and the natural hormone.

Oh, by the way, Monsanto sued a dairy for labeling its milk "No Artificial Growth Hormones," claiming that it had no right to let customers know if its milk contains genetically engineered hormones. No matter that rBGH has been banned in every industrialized nation in the world...except us...the US.

Another example: the deceptive "all-natural" statement, on American products is the second most used claim in advertising. Legally, the all-natural assertion is left wide open by the FDA to mean anything or everything. The term "all natural" can mean anything, includes derivatives, genetically modified organisms or synthesized ingredients.[lxxx]

Another example: Congressional Bills (April, 2013) HB 875, HB 814, SF 425 and soon HB 759 are designed to be the death of organic farming and independent farming in America. They merge the USDA and FDA, two corrupt agencies, into one centralized agency with more power and influence.[lxxxi]

"The thing that bugs me is that people think the FDA is protecting them. It isn't. What the FDA is doing and what the public thinks its doing are as different as night and day." Quote by Dr. Herbet Ley, Commissioner of FDA (San Francisco Chronicle, 1-2-70)[lxxxii]

Another example: The FDA instituted HACCP (Hazard Analysis Critical Control Point), wiped out 72 local meat processors in just the state of Kansas. Carol Tucker was in charge for this while working at Monsanto whom also showed no problems with e-coli, while creating a vastly more centralized "industrial" (corporate) system. Under this new system inspections fell and food safety problems increased.

"HACCP [was] the keystone of President Clinton's globalization strategy to restrict the ability of Congress and of citizens at risk of health to make safety a political, or policy issue." FDA's HACCP made "feces" an approved part of the American diet."[lxxxiii]

Another example: American cherry growers received money from the USDA for an independent study that showed cherries were potentially ten times stronger than an aspirin and ibuprofen for controlling pain. The FDA forbade them from publicizing these results.[lxxxiv]

The examples of both the USDA and FDA favoring industry to the literal destruction of safer small farmers or small businesses are endless. The USDA wiped out black farmers through massive discrimination and the FDA attacked and eliminated natural health practitioners.   The USDA and FDA have worked to eliminate all non-corporate, less expensive, more healthy, green, and local alternatives to industrial food and industrial drugs under the aegis of food or drug safety, while passing and protecting and promoting dangerous corporate products such as rBGH and Vioxx.[lxxxv]

"It is a crime for a farmer to save seeds for the next year's planting."[lxxxvi]

Monsanto has a revolving door into the FDA, USDA and EPA.[lxxxvii]

Another example: There was a campaign in 2011 to sign a petition to label modified foods. Almost a million signatures and fifty five politicians signed the petition called "Just Label It", however the FDA disallowed it because they signed the same single petition so it was just counted as…yep…one. Regretfully, the petition was deemed not relevant.[lxxxviii]

Reality

95% of the Universe exists beyond our perception of reality. Only the small portion of our world that falls within the visible light spectrum and audible sounds is detectable by man. Even our most sensitive instruments cannot increase this by very much. What is this matter that we cannot detect and still makes up most of our universe? Scientists call it "Dark Matter" because they cannot explain it, but know it is there.

Dark matter

Dark matter is hypothetical matter that is undetectable by its emitted radiation, but its presence can be inferred from gravitational effects on visible matter. Dark matter and dark energy account for 95% of the mass of the observable universe. It affects the rotational speeds
of galaxies, orbital velocities of galaxies in clusters, gravitational lensing of background objects and the temperature distribution of gas in the universe's galaxies.

Okay. Wrapping up with a conclusion:

We could discuss things more; however, we need to move on to our intended subject material. The point we made in this chapter is that everything is not always the way it seems. We at least must keep our minds open and approach matters with as much good scientific information as possible. Salt has been an emotional issue based on poor or no good science. I need to explain.

Next chapter is on salt.

Chapter Three – The War on Salt

Time to End War on the Good Salt— we do need Salt.

Okay, you have heard a lot about the bad things of salt. Here in this chapter we are going to take a small look at true good science that supports an *opposing view*. I am asking that you keep an open mind and more importantly, educate yourself...constantly. The story is not all one-sided. You need to know that we have been over impressionalized by drama. Calm down and read, then move to the next chapter.

Salt (sodium chloride) is the most ubiquitous food ingredient consumed by humankind. It is a nutrient that is essential to life and good health and has always been the predominantly useful ion in extracellular body fluid for all multi-cellular animals.

It's no secret that sodium has replaced carbohydrates and fat as the new *scapegoat* for all our woes of health issues. New product marketing strategies boast "low sodium" or "reduced sodium" labels, and many brands are replacing sodium altogether with potentially harmful potassium chloride. Legislation in cities and states is being passed to limit salt in our food. The salt police are out. The zealous drive by politicians to limit our salt intake has little basis in science.

The Mistaken Rational to Demonize Salt [Beginning of]

Worries escalated over salt in the 1970s when Brookhaven National Laboratory's Lewis Dahl claimed he had "unequivocal" evidence that salt causes hypertension: he induced high blood pressure in rats by feeding them the human equivalent of 500 grams of sodium (refined) per day. Today the average American consumes 3.4 grams of sodium, or 8.5 grams of salt, a day. Dahl believed he discovered population trends that continue to be cited as strong evidence of a link between salt intake and high blood pressure. People living in countries with high salt consumption—such as Japan—also tend to have high blood pressure and more strokes. But as a paper pointed out several years later in the *American Journal of Hypertension,* scientists had *found little such associations* when they compared sodium intake *within* populations, which *suggested that genetics or other cultural factors might be the culprit*. Nevertheless, in 1977 the U.S. Senate's Select Committee on Nutrition and Human Needs released a report recommending that Americans cut their salt intake by 50 to 85 percent, based largely on Dahl's work! Incredible!

War on Salt

The issue here is that the Government agencies are against all forms of salt, a broad-brush condemnation designed more for media sound bites than to truly advance the cause of health.[lxxxix]

Bloomberg....

Twenty-one U.S. companies have reduced salt content in pre-packaged or restaurant foods as part of an effort led by New York Mayor Bloomberg to slash the amount of sodium in the national diet.[xc]

> Good news. The bad salt used by processors is the unhealthy salt we will talk about later. Some of its reduction is welcome.

Bad news. The 21 companies participating in the salt reduction just replace the *bad* salt with *poison*! "Light Salt" and "Salt Replacements" are true killers.
Ugly news. No one is promoting using a healthy salt. Not one mention of it. Even if the salt is removed, consumers will add their own concoctions of bad salt. We will discuss this in detail later in the book.

Centers for Disease Control and Prevention

Centers for Disease Control and Prevention (CDC) has suggest salt is so bad for us, that curbing salt is as important as quitting smoking! This government agency was led by Dr Kimberly Lindsey whom left his mark by being involved with directives to fluoridate your water, created bogus swine flu propaganda and other fraud.[xci]

"And yet," writes Gary Taubes, a *Robert Wood Johnson Foundation* Independent Investigator in Health Policy Research and the author of *Why We Get Fat*, in *The New York Times*, "this eat-less-salt argument has been surprisingly controversial - and difficult to defend. Not because the food industry opposes it, but because the actual evidence to support it has always been so weak."

As early as 1998, when he first began researching the supposedly ill-effects of salt , Taubes said "many medical journal editors and public health administrators admitted that, even after 25 years' of an anti-sodium campaign, the evidence to support the claim was flimsy at best." [xcii]

While the mantra has always been more salt means more medical problems, Taubes said some research published within the past few years actually indicates that eating *less* salt contributes to dying prematurely.

"Put simply, the possibility has been raised that if we were to eat as little salt as the USDA and the CDC recommend, we'd be harming rather than helping ourselves," he wrote.

Salt Intake Research has Historically Been Inconclusive

The eat-less-salt recommendation, Taubes says, came from two bodies of research prior to 1972, when the *National Institutes of Health* (NIH) introduced the *National High Blood Pressure Education Program* to help prevent hypertension. No conclusive studies had been done before then, but the two bodies of research, while inconclusive, at least seemed to support the possibility that too much salt *could* cause health problems. So the government's "experts" ran with that.[xciii]

Today, the government agencies primarily rely on a single 30-day study called the *DASH-Sodium* study, conducted in 2001. "It suggested that eating significantly less salt would modestly lower blood pressure; it said nothing about whether this would reduce hypertension, prevent heart disease or lengthen life," Taubes says, pointing out that other analyses and meta-analyses of various salt-intake researches has produced more of the same: *inconclusive data*.

"This attitude that studies that go against prevailing beliefs should be ignored on the basis that, well, they go against prevailing beliefs, has been the norm for the anti-salt campaign for decades. Maybe now the prevailing beliefs should be changed," Taubes wrote.[xciv]

For years we've been told to lower our salt intake for our health. Individuals at risk for heart attack are especially admonished to drop their salt intake as low as possible. As it turns out, this seemingly harmless recommendation is actually putting us at a higher risk for conditions like heart disease and stroke. Although salt has been construed as a vital substance responsible for ruining our heart health, new research says too little salt may be just as harmful as too much.

SALT Consumption and Life Expectancy

Average life expectancy is sometimes considered a measure of the overall health of a population. Comparing the data on average sodium consumption in thirty-two countries around the world[34] with life expectancy results and if we take the top 20 percent with greatest life expectancy, their sodium intake is 1.75 teaspoons of salt per day. I am not claiming a direct cause-and-effect relationship between sodium intake and lifespan; however the data does demonstrate the *compatibility* between life expectancy and levels of sodium intake. [xcv]

Cochrane Collaboration; Studies Debunk Salt Risk

The "Cochrane Collaboration" is a large group of academics who publish systematic reviews of medical research papers regarding the effectiveness of various health care interventions, conventional and unconventional. The Cochrane Collaboration publishes these reviews and can be accessed on line by doctors and individuals. Following is a review by this science team.

Cochrane Summary: "We are commonly advised to cut down on salt. Analyses of studies with a duration of 2 to 4 weeks or longer were performed. Low salt diets reduced systolic blood pressure by 1% in people with normal blood pressure and by 3.5% in people with elevated blood pressure. The effect was similar in trials of 4 weeks or longer.

There were *increases* in some *hormones and lipids* which could be *harmful* (*due to low salt*) if persistent over time. However, the studies were not designed to measure long-term health effects. Therefore we do not know if low salt diets improve or worsen health outcomes.[xcvi]

Cochrane Summary of INTERSALT Study: "Scientific tools have become much more precise since then, but the correlation between salt intake and poor health has remained tenuous. "INTERSALT", a large study published in 1988, compared sodium intake with blood pressure in subjects from 52 international research centers and found no relationship between sodium intake and the prevalence of hypertension. In fact, the population that ate the most salt, about 14 grams a day, had a lower median blood pressure than the population that ate the least, about 7.2 grams a day. In 2004 the Cochrane Collaboration, an international, independent, not-for-profit health care research organization funded in part by the U.S. Department of Health and Human Services, published a review of 11 salt-reduction trials. Over the long-term, low-salt diets, compared to normal diets, *decreased systolic blood pressure* (the top number in the blood pressure ratio) in healthy people by 1.1 millimeters of mercury (mmHg) and diastolic blood pressure (the bottom number) by 0.6 mmHg. That is like going from 120/80 to 119/79. The review concluded that "intensive interventions, unsuited to primary care or population prevention programs, provide only minimal reductions in blood pressure during long-term trials." A 2003 Cochrane review of 57 shorter-term trials similarly concluded that "there is little evidence for long-term benefit from reducing salt intake."

New York Times 5/3/11: Published an article of a study that showed Low-Salt diets increased the risk of death from heart attacks and strokes. [xcvii]

NCBI (Government agency-National Center for Biotechnology Information) report: Meta-analysis shows that blood pressure was reduced systolic by 1.1 mm Hg and diastolic by .6mm Hg at 13 to 60 months. ...there are doubts about effects of sodium reduction on overall health.[xcviii]

Dr. Jeffrey R. Cutler- (Jul 1, 2008) (past president of the American Heart Association): Documented no health outcomes benefits of lower-sodium diets. Dr Jeffrey is the past president of the American Heart Association. He also lists 17 studies that show no health benefits for low sodium diets.[xcix]

Published in the *Daily Express* and the peer-reviewed *American Journal of Hypertension*: Article claimed that "salt is safe to eat", and that, after years of lecturing, the "health fascists" have been proved wrong.
This news is based on a systematic review that combined data from seven earlier studies looking at how reduced-salt diets affected the risk of cardiovascular disease (CVD), blood pressure and death.[c]

Vilifying Salt—Your Government at work

- Department of Agriculture's dietary guidelines still consider salt Public Enemy No. 1, coming before fats, sugars and alcohol.
- Centers for Disease Control and Prevention have suggested that reducing salt consumption is as critical to long-term health as quitting cigarettes.
- U.S.D.A., the Institute of Medicine, rely on results of a 30 day trial with 60 times normal intake
- C.D.C. , rely on results of a 30 day trial with 60 times normal intake
- N.I.H. — all essentially rely on the results from a 14-day trial of salt, with high sodium levels of 60 times the use by average American

Drummond Rennie, an editor for The Journal of the American Medical Association, has stated that the authorities pushing the eat-less-salt message had "made a commitment to salt education that goes way beyond the scientific facts."

For many years we have been told by medical experts and nutritionists alike that it is crucial for our health to cut down on sodium as much as possible. Statistical data _seems_ to show that a high dietary salt intake (primarily consisting in sodium chloride) can put people at risk of cardiovascular complications.

Consequently, salt has been vilified to such extent in the media that may people strive to remove it from their diets completely. Wiping salt out of the menu is not the right choice. The government and health community have chosen the wrong villain. The _type_ of salt, the _bad_ salt, we ingest is the real villain.

Government running amuck—

Put simply, the possibility has been raised that if we were to eat as little salt as the U.S.D.A. and the C.D.C. recommend, we'd be harming rather than helping ourselves.

Consuming salt causes hypertension is only a hypothesis. [ci] Dr Jeffrey Cutler past president of the American heart Association explains in an article that there are many factors not accounted for. He states "hypertension is a "sign "and simply reducing salt may not be a benefit over the side effects like reduced minerals, hormones and how the body must work with lower salinity. He lists 17 citations that scientifically show that a low salt diet has no benefits in this endnote-[cii]

In 1972, the National Institutes of Health introduced the National High Blood Pressure Education Program to help prevent hypertension released two pieces of research.

- One was the observation that populations that ate little salt had virtually no hypertension. (they did not eat a lot of things)

- The second was a strain of "salt-sensitive" rats that reliably developed hypertension on a high-salt diet. The catch was that "high salt" to these rats was <u>60 times more</u> than what the average American consumes.[ciii]

- Many conditions effect hypertension like genetics. Here is the Dahl report.[civ]

Eating less salt seemed to be the only available option at the time, short of losing weight. Although researchers quietly acknowledged that the data were "inconclusive and contradictory" or "inconsistent and contradictory" — two quotes from the cardiologist Jeremiah Stamler, a leading proponent of the eat-less-salt campaign, in 1967 and 1981 — publicly, the link between salt and blood pressure was upgraded from hypothesis to fact.

Four years ago, Italian researchers began publishing the results from a series of clinical trials, all of which reported that, among patients with heart failure, reducing salt consumption increased the risk of death.[cv]

The *Department of Agriculture*'s (USDA) dietary guidelines, released in January 2011 for the first time in six years, made the recommendation to slash one's intake of salt to no more than 2,300 milligrams daily, or about one tablespoon. The USDA further recommended that for people 51 and older, all African-Americans, and people who have hypertension, diabetes or chronic kidney disease (regardless of their age) cut their intake to about 1,500 milligrams![cvi] Another reference-[cvii]

Amazing! The body is 57% salt water. A human body (200 lbs) would have 200 grams or 200,000 mg of salt. You would lose 1,500 mg of salt blowing your nose! You would lose 1,000 mg crying at a funny story with friends!

In recent years there has been much publicity about the need to reduce salt consumption in societies where salt is added to many processed foods (Denton 1984, 584-7). Well, one thing is true...we do *not* want the bad salt from processed food. The salt used in processed food is the wrong king of salt to begin with, however, first things first. Following is a discussion on how too little salt is not good.

Let's start here:

It has been forgotten that some salt intake is absolutely necessary; that people need salt, sodium chloride, to survive. The chemical requirements of the human body demand that the salt concentration in the blood be kept constant. If the body does not get enough salt, a hormonal mechanism compensates by reducing the excretion of salt in the urine and sweat. But it cannot reduce this output to zero. On a completely salt-free diet the body steadily loses small amounts of salt via the kidneys and sweat glands. It then attempts to adjust this by accelerating its secretion of water, so that the blood's salt concentration can be maintained at the vital level. The result is a gradual desiccation of the body and finally *death*."

Contrary to science

The government and recent publications are attempting to change the human requirement for salt. This attempt to lower salt is quite contrary to science.

I. Harvard Medical School demonstrated that when healthy people were placed on a very low-salt diet a fifth of a teaspoon (1 gram) of salt per day, they developed insulin resistance within seven days. This causes your body to create more insulin and this is devastating to the body! Those placed on higher salt diets of 8 grams showed *no* such effect. Low-salt intake contributes to diabetes and cardiovascular disease.[cviii]

II.   In a series of three analyses of consecutive National Health and Nutrition Examination Surveys (NHANES), researchers were unable to demonstrate any survival advantage resulting from low-sodium diets; on the contrary, a modest relationship between *increased all-cause mortality* and low-sodium diets was observed. [cix]

III.   Observed associations of lower sodium with higher mortality were modest and mostly not statistically significant.[cx]

IV.   A recent study conducted to examine the health outcomes related to salt intake demonstrated that lower sodium excretion was associated with an *increased* risk of cardiovascular death, while higher sodium excretion did not correspond with increased risk of hypertension or cardiovascular disease complications.[cxi]

V.   Another meta-analysis of one hundred sixty-seven studies by Graudal and co-workers confirmed and expanded upon previous reports that significant dietary sodium restriction from greater than or equal to 1.5 teaspoons, down to 1.2 teaspoons per day resulted in lowering the blood pressure by only a few points with high BP and almost none for a person with normal BP. [cxii]

VI.   However, the meta-analysis went further to confirm and quantified the unfavorable impacts that sodium restriction had on several other risk factors for cardiovascular disease. These included significant increases in renin, aldosterone, catecholamines (adrenaline, noradrenalin) and lipids (cholesterol and triglycerides).[cxiii]

For decades, policy makers have tried and failed to get Americans to eat less salt. In April 2010 the *Institute of Medicine* urged the *U.S. Food and Drug Administration* to regulate the amount of salt that food manufacturers put into products; New York City Mayor Michael Bloomberg has already convinced 16 companies to do so voluntarily. But if the U.S. does conquer salt, what will we gain? Bland french fries, for sure. But a healthy nation? Not necessarily.

High Blood Pressure and Salt

A meta-analysis of seven studies involving a total of 6,250 subjects in the *American Journal of Hypertension* found no strong evidence that cutting salt intake reduces the risk for heart attacks, strokes or death in people with normal or high blood pressure. European researchers publishing in the *Journal of the American Medical Association* reported that the less sodium that the study subjects excreted in their urine—an excellent measure of prior consumption—the *greater* their risk was of "dying" from heart disease. These findings call into question the common wisdom that excess salt is bad for you, but the evidence linking salt to heart disease has always been tenuous.[cxiv]

Salt is currently considered a leading culprit for high blood pressure and other health problems. This is based on the premise that a high sodium intake creates high blood pressure, which can lead to heart attacks, arterial, and kidney problems. Salt is high in sodium. But sodium is essential for proper absorption of other major nutrients and functioning of nerves and muscles, as well as being necessary for balancing water and minerals in the body.[cxv]

Worries regarding salt escalated in the 1970s when Brookhaven National Laboratory's Lewis Dahl claimed that he had "unequivocal" evidence that salt causes hypertension: he induced high blood pressure in rats by feeding them the human equivalent of 500 grams of sodium a day. (Today the average American consumes 3.4 grams of sodium, or 8.5 grams of salt, a day.)

Dahl also discovered population trends that continue to be cited as strong evidence of a link between salt intake and high blood pressure. People living in countries with a high salt consumption—such as Japan—also tend to have high blood pressure and more strokes. But as a paper pointed out several years later in the *American Journal of Hypertension,* scientists had little luck finding such associations when they compared sodium intakes *within* populations, which suggested that genetics or other cultural factors might be the culprit. Nevertheless, in 1977 the U.S. Senate's Select Committee on Nutrition and Human Needs released a report recommending that Americans cut their salt intake by 50 to 85 percent, based largely on Dahl's work.[cxvi]

Chapter Four— Goodness of Salt

Miracle of Salt

An abundance of the minerals and elements in unrefined salt are as synonymous with life today as they were a billion years ago. Even single cells require salt and minerals. Lack of them is synonymous with birth defects, organ failure, decay, diseases, premature aging and death at a young age.

Science and medicine have tried to define the precise roles of salt in the healthy and diseased human organism. Blood, sweat, and tears all contain salt, and both the skin and the eyes are protected from infectious germs by the anti-bacterial effect of salt. When salt is added to a liquid, particles with opposite charges are formed: a positively charged sodium ion and a negatively charged chloride ion. This is the basis of osmosis, which regulates fluid pressure within living cells and protects the body against excessive water loss (as in diarrhea or with heavy sweating). Sodium and chloride ions, as well as potassium ions, create a measurable difference in potential across cell membranes. This ensures that the fluid inside living cells remains separate from that outside. Thus, although the human body consists mainly of water, our "inner ocean" does not flow away or evaporate. Sodium ions create a high pressure of liquid in the kidneys and thus regulate their metabolic function. Water is extracted through the renal drainage system. The body thus loses a minimal amount of essential water. While approximately 1500 liters of blood pass through the kidneys daily, only about 1.5 liters of liquid leave the body as urine.

Salt Intake is Vital

Salt is a vital substance for the survival of all living creatures, particularly humans. Water and salt play a major role in regulating the water content of the body. Water itself regulates the water content of the interior of the cell by working its way into all of the cells it reaches. Water is required to cleanse and extract the toxic wastes of cell metabolisms.

Salt plays an important role in regulating the balance for water in cells by forcing some water to stay outside the cells. There are two oceans of water in the body; one ocean is held inside the cells of the body, and the other ocean is held outside the cells. Good health depends on a most delicate balance between the volume of these oceans, and this balance is achieved in large part by salt. Later we will discuss the nutritional salt that is natural unrefined and much easier for the body to incorporate.

Insufficient Water

When water is available to get inside the cells freely, it is filtered from the outside salty ocean and injected into the cells. The cells are being overworked despite their water shortage. This is the reason why in dehydration (not drinking enough water) we develop edema [*or dropsy, an accumulation of water in the body*] and retain water. The design of our bodies is such that the extent of the ocean of water outside the cells is expanded to have the extra water available for filtration and emergency injection into vital cells. The brain commands an increase in sodium and water retention by the kidneys.

Initially, the process of water filtration and its delivery into the cells is more efficient at night when the body is horizontal. The collected water, mostly pooling in the legs, does not have to fight the force of gravity to get into the blood circulation. If reliance of this process of emergency hydration of some cells continues for long, the lungs begin to get waterlogged at night, and breathing becomes difficult. The person needs more pillows to sit upright to sleep. In this case, this condition is the consequence of dehydration.

Hydration is of course needed; however, you might overload the system by drinking too much water at the beginning. Increases in water intake must be slow and spread out until urine production begins to increase at the same rate that you drink water.

When we drink enough water to pass clear urine, we also pass out a lot of the salt that was held back. In this situation this is how we can get rid of edema fluid in the body; by drinking more water. Not diuretics, but more water! In people who have an extensive edema and show signs of their heart beginning to have irregular or very rapid beats with least effort, the increase in water intake should be gradual and spaced out, but not withheld from the body. Naturally, salt intake should be limited for two or three days because the body is still in an overdrive mode to retain it. Once the edema has cleared up, salt should not be withheld from the body.

Salt Functions

Salt has many other functions than just regulating the water content of the body. Here are some of the more vital functions of salt in the body:

1. Salt is important for water balance regulation and for fluid distribution on either side of the cell walls.
2. Salt is very important to the proper function of the adrenal glands.
3. Salt is very effective in stabilizing irregular heartbeats and, contrary to the misconception that it automatically causes high blood pressure, it is actually essential for the regulation of blood pressure – in conjunction with water. Naturally the proportions are critical as are all chemicals in our body. Cells and systems within us miraculously keep over fifty chemicals in balance.
4. Salt is vital to the extraction of excess acidity from the cells in the body, particularly the brain cells.
5. Salt is vital for balancing the sugar levels in the blood; a needed element in diabetics.
6. Salt is vital for the generation of hydroelectric energy in cells in the body. It is used for local power generation at the sites of energy need by the cells.
7. Salt is vital to the nerve cells' communication and information processing all the time that the brain cells work, from the moment of conception to death.
8. Salt is vital for absorption of food particles through the intestinal tract.
9. Salt is vital for the clearance of the lungs of mucus plugs and sticky phlegm, particularly in asthma and cystic fibrosis.
10. Salt is vital for clearing up catarrh and congestion of the sinuses.
11. Salt is a strong natural antihistamine.
12. Salt is essential for the prevention of muscle cramps.

13. Salt is vital to prevent excess saliva production to the point that it flows out of the mouth during sleep. Needing to constantly mop up excess saliva indicates salt shortage.
14. Salt is absolutely vital to making the structure of bones firm. Osteoporosis, in a major way, is a result of salt and water shortage in the body.
15. Salt is vital for sleep regulation. It is a natural hypnotic.
16. Salt is a vitally needed element in the treatment of diabetics.
17. Salt on the tongue will stop persistent dry coughs.
18. Salt is vital for the prevention of gout and gouty arthritis.
19. Salt is vital for maintaining sexuality and libido.
20. Salt is vital for preventing varicose veins and spider veins on the legs and thighs.
21. Salt is vital to the communication and information processing nerve cells the entire time that the brain cells work – from the moment of conception to death.
22. Salt is vital for reducing a double chin. When the body is short of salt, it means the body really is short of water. The salivary glands sense the salt shortage and are obliged to produce more saliva to lubricate the act of chewing and swallowing and also to supply the stomach with water that it needs for breaking down foods. Circulation to the salivary glands increases and the blood vessels become "leaky" in order to supply the glands with water to manufacture saliva. The "leakiness" spills beyond the area of the glands themselves, causing increased bulk under the skin of the chin, the cheeks and into the neck.
23. Sea salt contains about 80 mineral elements that the body needs. Some of these elements are needed in trace amounts. Unrefined sea salt is a better choice of salt than other types of salt on the market. Ordinary table salt that is bought in the super markets has been stripped of its companion elements and contains additive elements such

as aluminum silicate to keep it powdery and porous. Aluminum is a very toxic element in our nervous system. It is implicated as one of the primary causes of Alzheimer's disease.

24. Twenty-seven percent of the body's salt is in the bones. Osteoporosis results when the body needs more salt and takes it from the body. Is it not obvious what happens to the bones when we're deficient in salt or water or both?[cxvii]

Anti-bacterial Effect of Salt

- Salt particles kill the bacteria and fungi on mucous membranes while at the same time strengthening the immune system.
- Salt kills the germs on the food you dash salt onto.
- Salt water kills internal parasites and bacteria.[cxviii]
- Salt also reduces the swelling of respiratory mucous membranes: sneezing stops, nose unblocks, and the cold reduces or disappears.
- Salt aerosol has a positive effect on the nervous system and the ability to reduce stress.
- Salt therapy also relieves many skin problems by killing bacteria on the skin.
- Salt is a good food preservative and bacteria killer. Actually, if high levels of salt are removed from processed food, other preservatives and bacterial killing agents would need to be added.

Nerve Health

Salt is "fuel" for nerves. Streams of positively and negatively charged ions send impulses to nerve fibers. A muscle cell will only contract if an impulse reaches it. Nerve impulses are partly controlled by coordinated changes in charged particles where the polarization of the neuron's membrane has sodium on the outside and potassium on the inside. The outside of the cell is positive while the electrical charge on the inside of the membrane is negative. The outside contains excess sodium ions and the inside cell contains excess potassium.[cxix]

Natural sea salt is the best source of sodium and potassium for the body.

Salt for Bones

Salt is an adamant requirement for the proper PH of the body. Calcium, identical to that in rocks and mollusk shells, keeps bones rigid and strong. Calcium salts in bone are in dynamic balance with blood and tissues and provide a ready source of alkalinizing material to counter an acid challenge. But, like money, the supply is not infinite. While we spend our first eighteen or so years growing and building bone, we spend the rest of our lives tearing it back down, a process regulated by body PH. The chronic mild metabolic acidosis engendered by our diet worsens as we age, starting in our teens and continuing through the eighth decade. 5.6 acidic pH food pulls calcium carbonate and calcium phosphate from bone to maintain the body pH of 7.4. The problem comes when you habitually ingest acids in the diet, then draw on calcium stores over and over and over again to neutralize these acids.

Though bones have a lot of stored calcium, the supply is not inexhaustible. Bones will eventually become demineralized, i.e., depleted of calcium. That's when osteopenia (mild demineralization) and osteoporosis (severe), frailty, and fractures develop. Incidentally, taking calcium supplements has little effective at reversing bone loss.[cxx] Sodium plays an important role in keeping the body alkaline and from losing bone mass.

People who eat high-salt diets from highly processed food are consuming a lot of "salt. This very processed salt is more difficult for your body to use than unrefined, natural salt. When sodium is lost in the urine, it drags calcium with it. Too much calcium in your urine may cause kidney stones and contribute to bone loss. Calcium and sodium (salt) are handled in a similar way by the kidney. Anything that causes increased sodium (salt) excretion by the kidney will, *en passant*, cause increased calcium excretion by the kidney.

Iodized Salt

Iodine but not in salt! Iodine is an important element that is essential for the body and particularly its thyroid gland. Iodine also helps balance the process of fats stored in the body, and calcium utilization. Too little iodine and we get goiter.

Problem: The iodine in iodized salt is inorganic and not natural. Our body cannot properly assimilate it. Iodized salt is simply table salt, made inorganic by extreme over processing, with added inorganic potassium iodide. Table salt has been altered until it is rendered into a lifeless compound that the body cannot assimilate easily nor eliminate easily. Adding another insult does not make the consumption any better.

Iodide is unstable and can easily break down. To counter this, sodium ferrocyanide is added along with Dextrose to help stabilize the Iodide.

At high temperatures however, such as searing meats and baking, some of the Iodide can still be broken down and oxidized, turning it into Iodine, which gives off an acrid smell or flavor.

Suggestion: Good and safe sources of iodine to prevent goiter are sea algae of all kinds (especially kelp), yogurt and eggs.

Chapter Five— Chemistry of Salt

Glad you are still here. It's nice having a chat with you, but before we delve into a little more intense reading...let's take a quick stroll down by the sea and go over *where* salt comes from and *how* it is made. Ah, sometimes I can smell the ocean wafting in the wind here.

> I walked by the seashore with sand under my feet,
> The wind blew on my face a scent so sweet.
> Looking out yonder my eyes saw endless blue
> Never could I imagine its depth or mystery... no clue!
> The tides keep rising, bringing new ones near,
> Washing the sandy beaches with waters sparkling clear,
> What do these waves carry? What jewel is in its hand?
> This brinish potion hangs in the air and over sand.
> I walk away home, but these memories linger on...
> Long after, my footprints on the sands are gone

JS

Creation of Salt

Rain falls over the earth and runs to the streams and rivers washing with it small quantities of salt and deposit it a little at a time into our lakes, seas and oceans. Oceans and seas that have no outlet to allow the water to move on must keep their silt, sand, rock particles, minerals and dissolved salts. Evaporation of water dissolves into the atmosphere leaving the mineral deposits to build up in the ocean. Over time, the water becomes more concentrated with the minerals. Since salt dissolves so well in water, it becomes salty. The heavier non dissolving minerals and silt sink to the bottom.

All salt is collected from this procedure. Salt beds of gathered and deposited salt are found where ancient water has dried up. Perhaps due to tectonic plates changing the elevation or maybe the water source like rivers has stopped, sometimes these ancient salt deposits are buried deep into the ground; however, that is how they are formed. Whether the salt was mined under the earth, taken from salt flats or more likely, trapping salty water and allowing the water to evaporate, the formation has not changed.

More on salt *creation.* Thank you for journeying a little further into the amazing story of where salt really comes from.

The marvel of *salt creation* is that the elements of sodium and chloride do not mix. Salt is not a natural element and does not form in nature alone. True!

You can place the element sodium (a metal) on the element chloride and *nothing* will happen. They will sit there touching but no reaction will occur and no salt (sodium chloride) will form. Add even a small amount of water and you have a very violent fire and explosion!

Salt is actually a general name but the most commonly known is sodium chloride, or table salt, is a compound formed by the chemical reaction of an acid with a base. During this reaction, the acid and base are neutralized producing 1) salt, 2) water and 3) heat. Sodium chloride is distributed throughout nature as deposits on land and then created by the evaporation of ancient seas. It is also dissolved in the oceans. [cxxi]

Creation—almost done:

In terms of chemistry, a salt can be any compound formed by the reaction of an acid with a base. Energy, in the form of heat, is given off during this neutralization reaction so it is said to be exothermic. The most common salt, **sodium chloride** (NaCl), is a product of the reaction between **hydrochloric acid** (HCl) and the base **sodium hydroxide** (NaOH). In this reaction, positively charged hydrogen ions (H+) from the acid are attracted to negatively charged hydroxyl ions (OH-) from the base. These ions combine and form water. After the water forms, the sodium and chlorine ions remain dissolved and the acid and base are said to be neutralized. Solid salt is formed when the water evaporates and the negatively charged chlorine ions combine with the positively charged sodium ions.[cxxii]

HCL (hydrochloric acid) + NaOH (sodium hydroxide) = NACL (salt) + H20 + heat

Let's repeat the creation of salt:

Sodium chloride (NaCl) is a product of the reaction between hydrochloric acid (HCl) and the base-sodium hydroxide (NaOH).[cxxiii]

A point needs to be made here while we are discussing how salt forms. Since NaCl is a derivative of sodium hydroxide the inverse can be true. Sodium hydroxide, known as *lye* is a highly alkali chemical base and is used in detergents and as drain cleaners.

Sodium Balance

Not too much and not too little. Though the amount deemed as excessive sodium in the body varies from one person to the next, the amount is much higher than what is currently being purported. Factors like sweating obviously effect the requirements for salt, however there are many hidden ones also. The presence of fluoride in the blood blocks sodium as insulin and other chemicals. The health of the veins and cells effects salt intake needs.

Salt insufficiencies can result from too little salt quickly. Playing around with dietary salt reductions can create a cascade of health problems. To enlighten some awareness on the interactions of how our bodies work with salt lets walk through some activity on a cellular level.

After a salty meal, the movement of sodium from the blood into the interstitium is delayed by the significant buffering capacity of the endothelial glycocalyx. Sodium will reversibly bind to (dissociate) from the endothelial glycocalyx binding sites where it can be readily excreted via the kidneys. After disenfranchising from the glycocalyx, sodium gains direct access to the "unprotected" epithelial sodium channels. So, in addition to the para-cellular transport route (i.e. sodium transport *between* endothelial cells along its chemical gradient), sodium uses the transcellular pathway for entering the large extracellular space (about 30% of body weight). There, sodium is bound to the extracellular matrix.

Excessive *(a variant term that likely is much higher you're your current intake)* sodium intake over time will damage the endothelial glycocalyx and lead to a decrease in its sodium buffering capacity because of the loss of negatively charged heparin sulphate residues. Abusive amounts of salt lead to the delay of sodium excretion. After the ingested sodium has spread throughout the body, the plasma sodium is diluted and less concentrated. Then, the sodium diffuses back into the vascular bed (along its chemical gradient directed from interstitium to blood) and will finally be excreted by the kidneys. This "detour of sodium", (by going through the whole organism) delays renal sodium excretion. In the meantime, a sodium load from the next salty meal may arrive and so on, leading over time to sodium accumulation in the organism.[cxxiv]

Salt—the Ion Compound

Sodium chloride, [salt] is an ionic compound with the formula NaCl, representing equal proportions of sodium and chloride.

Sodium has one electron in its outer shell. Chlorine has seven electrons in its outer shell. When they react sodium loses an electron to become a sodium ion with a charge of +1. Chlorine atoms accept an electron to become chloride ions with a charge of -1. After reacting, the sodium and chloride ions have a full outer shell of electrons. A full outer shell of electrons is a stable arrangement. The inert gasses in group 18 of the periodic table all have a full outer shell and characteristically are very unreactive (they don't easily combine with other elements). The new substance sodium chloride is made up of ions. Compounds that have ions bonded together are called ionic. The ions are held together by powerful electrostatic forces. Ionic compounds typically have high melting and boiling points as a consequence.

Sodium (Na) is a highly reactive metallic element with an atomic number 11 with only one stable isotope. Chlorine (Cl) with the atomic number of 17 is a powerful oxidant and is used in bleaching and disinfectants. When you combine these two elements you end up with NaCl which it is salt .You combine them by transferring one electron from Na to Cl and then they become stable by having eight electrons each.

Note: If you were to eat sodium (a soft metal) or inhale chlorine gas, there is a strong chance it would kill you. However, when these two elements combine, you can safely sprinkle the resulting compound on your popcorn. The reason is that the elements have *neutralized* by sharing their electrons. [cxxv]

Facts—Sodium

Sodium unlike the other alkali metals has a higher density.
Sodium has 13 isotopes, and having only one stable isotope.
Sodium cannot exist freely in nature when brought in contact with air. It readily oxidizes in air and forms oxides
Sodium explodes when brought into contact with water.
Sodium is the fourth abundant element.
Storing sodium is extremely difficult. This is done by immersing into liquid hydrocarbon—like kerosene.
Sodium is used in steel alloys.
Sodium gas is used in street light bulbs.
Sodium is a significant ingredient in soap. [cxxvi]

Halite

Halite, commonly known as rock salt, is the mineral form of sodium chloride (NaCl). Halite forms isometric crystals. This hard rock mineral is typically colorless or white, but may also be light blue, dark blue, purple, pink, red, orange, yellow or gray depending on the amount and type of impurities. It commonly occurs with other evaporite deposit minerals such as several of the sulfates, halides, and borates.

Halite occurs in vast beds of sedimentary evaporite minerals *that result from the drying up of enclosed lakes, playas, and seas.* Salt beds may be hundreds of meters thick and underlie broad areas. In the United States and Canada extensive underground beds extend from the Appalachian basin of western New York through parts of Ontario and under much of the Michigan Basin. Other deposits are in Ohio, Kansas, New Mexico, Nova Scotia and Saskatchewan. The Khewra salt mine is a massive deposit of halite near Islamabad, Pakistan.

Unusual, purple, fibrous vein filling halite is found in France and a few other localities. Halite crystals termed *hopper crystals* appear to be "skeletons" of the typical cubes, with the edges present and stair step depressions on, or rather in, each crystal face. In a rapidly crystallizing environment, the edges of the cubes simply grow faster than the centers. Halite crystals form very quickly in some rapidly evaporating lakes resulting in modern artifacts with a coating or encrustation of halite crystals. *Halite flowers* are rare stalactites of curling fibers of halite that are found in certain arid caves of Australia's Nullarbor Plain. Halite stalactites and encrustations are also reported in the Quincy native copper mine of Hancock, Michigan.[cxxvii]

Solid sodium chloride

Cubic crystal system

In solid sodium chloride, each ion is surrounded by six ions of the opposite charge as expected on electrostatic grounds. The surrounding ions are located at the vertices of a regular octahedron. In the language of close-packing, the larger chloride ions are arranged in a cubic array whereas the smaller sodium ions fill all the cubic gaps (octahedral voids) between them. This same basic structure is found in many other compounds and is commonly known as the halite or rock-salt crystal structure. It can be represented as a face-centered cubic (fcc) lattice with a two-atom basis or as two interpenetrating face centered cubic lattices. The first atom is located at each lattice point, and the second atom is located half way between lattice points along the fcc unit cell edge.

Thermal conductivity of NaCl as a function of temperature has a maximum of 2.03 W/ (cm K) at 8 K and decreases to 0.069 at 314 K (41 °C). It also decreases with doping.

Aqueous solutions

The attraction between the Na^+ and Cl^- ions in the solid is so strong that only highly polar solvents like water dissolve NaCl well.

When dissolved in water, the sodium chloride framework disintegrates as the Na^+ and Cl^- ions become surrounded by the polar water molecules. These solutions consist of metal aquo complex with the formula $[Na(H_2O)_8]^+$, with the Na-O distance of 250 pm. The chloride ions are also strongly solvated, each being surrounded by an average of 6 molecules of water. Solutions of sodium chloride have very different properties from pure water. The freezing point is −21.12 °C for 23.31 wt% of salt, and the boiling point of saturated salt solution is near 108.7 °C. (227.7 F) From cold solutions, salt crystallizes as the dehydrate $NaCl \cdot 2H_2O$.[cxxviii]

Monosodium glutamate

Monosodium glutamate has been used for over five decades as a flavor intensifier. If you are reading your labels in the store you are seeing this chemical a lot.

It acts as a neurotransmitter, increasing the sensitive response of the sense of taste, acting in the transmission of electric impulses throughout the nervous system. Normally it is used in precooked meals such as soups, sauces, meats, tinned food, dressings, etc.

The problem is that, many studies have seriously questioned the safety of monosodium glutamate. Japanese researchers have associated it with the loss of vision and blindness in the long term.

A clinical investigation of the Universidad Complutense de Madrid has shown that monosodium glutamate intake considerably increases the appetite, increasing the risk of obesity.

Synthetically, monosodium glutamate may produce: muscle contractions in the face and chest, palpitations, asthma attack, and headaches, sterility, obesity, and muscle stiffness in the neck and jaws, brain cells degeneration, gastric problems, stiffness and/or weakness in the extremities, blurry vision, dizziness, thorax oppression, hot and tingling sensation, numbness and face blushing.

Chapter Six—Low Salt Symptoms

Sodium is essential for the body's daily functions. It helps in maintaining the electrolyte balance of our bodies and plays a vital role in the working of nerves and muscles.

Salt (sodium chloride) is .4% of the human body by weight and human plasma contains 0.9 percent sodium chloride. It has been determined that the minimum daily intake of salt is 1.5 teaspoons (8 grams).[cxxix]

The normal blood sodium level is 135-145 milliequivalents per liter (mEg/l). Anything lower than 135 mEg/l indicates abnormally low sodium levels and is known clinically as hyponatremia. The main complication of hyponatremia is the swelling it causes in the body. Yes, too *little* salt causes swelling. A major _drop_ in sodium level causes cells to take in excess water and then _swell_ up. This is especially serious in the brain cells.

A recent study published in the *Journal of the American Medical Association*, explains just how important salt is to our body functioning properly.

A meta-analysis of seven studies involving a total of 6,250 subjects in The *American Journal of Hypertension* found no strong evidence that cutting salt intake reduces the risk for heart attacks, strokes or death in people with normal or high blood pressure.

Low-sodium diets trigger a negative chain reaction in the body that increases the risk of diabetes, stroke, heart attack and heart disease.[cxxx]

To understand *why* a) salt is healthy and b) *why* it has the potential to be dangerous, let's look at what it can do for us. The primary biological role of sodium is to regulate blood volume and blood pressure by maintaining adequate body fluid levels. When the kidneys detect too little sodium levels in the body, they decrease sodium excretion. But when there is too much sodium, an anti-diuretic hormone kicks in and causes the body to retain water. The kidneys will then try to gradually release excess sodium and water through urine, thus bringing the body's fluid and sodium levels back within normal ranges.

Water and salts are also lost through excessive perspiration, associated with hot climates and physical effort. This can severely offset the body's internal regulating mechanisms, and adequate rehydration is advised as soon as possible. In normal conditions however, sodium regulation is left almost entirely up to the kidneys. Medical experts usually warn us that if for some reason, the kidneys are unable to excrete excess sodium; the increased blood volume will exert extra pressure on blood vessels and make the heart work harder.

Now, scientists say that <u>too little sodium is just as bad.</u> On the one hand, sodium deficiency can cause a range of problems, including headaches, nausea, fatigue and muscle cramps. On the other hand, commercially available foods contain large amounts it difficult for people on a traditional diet to control their sodium intake.

The research of Professors Martin J. O'Donnell and Salim Yusuf of McMaster University in Ontario, Canada now reveals if we have very low levels of sodium in urine, we are at risk of cardiovascular death and congestive heart failure. The study looked at 28,880 people, with an elevated risk of cardiovascular disease, over a seven-year period. Those who had low sodium intake, between 2 and 3 grams per day, also showed a 20 percent higher risk of cardiovascular-related death as well as hospitalization for congestive heart failure. [cxxxi]

Low blood sodium (hyponatremia) occurs when you have an abnormally low amount of sodium in your blood, or when you have too much water in your blood. Low blood sodium is common in older adults, especially those who are hospitalized or living in long-term care facilities.

Older adults usually become ill with hyponatremia due to age-related causes that affect the way the body handles the balance of sodium and water, such as:

- Taking certain medications, such as diuretics, antidepressants and pain medications
- Changes in kidney function, such as decreased kidney size or decreased blood flow through the kidneys
- Severe vomiting or diarrhea
- Liver failure (cirrhosis)
- Kidney failure
- Heart failure
- Having high levels of anti-diuretic hormone, which causes you to retain water
- Drinking too much water
- Urinating less frequently
- Underactive thyroid (hypothyroidism)
- Addison's disease, a condition affecting the adrenal gland.[cxxxii]

Symptoms that Indicate Hyponatremia [low salt]

This condition is diagnosed with blood tests and urine tests but have the following symptoms:

- Nausea or vomiting
- Confusion, confusion and disorientation (due to brain functions)
- Headache, lethargy and fatigue
- Loss of appetite
- Irritability
- Restlessness, seizures, muscle weakness, spasms and cramps
- Reduced consciousness and even possible coma

Hyponatremia can be severe when it occurs suddenly and rapidly due to the fast swelling effect on the brain. People considered more susceptible to hyponatremia are:

- Women in premenopausal stage
- Older people
- Those people with kidney problems
- People taking diuretics
- Patients placed on low-salt or no-salt diets
- Athletes performing intense physical activities
- People in extreme heat or hot weather

Mild cases are managed by changing diets, lifestyles and medicines. If it is caused by drinking too much water a patient would be asked to reduce their fluid intake and/or increase their salt intake. For most healthy people, the correct amount of water intake should produce pale yellow urine.

Fact is that there is a lot of confusion about the proper balance of sodium (salt) for our bodies. The signs and symptoms of hyponatremia vary. A doctor can confirm whether you have hyponatremia by a blood test. Also a urine analysis can indicate sodium loss. The problem arises when, even with a normal blood level sodium level of 135-145 meq/L, patients are placed on unrealistic and unsustainable low salt diets.

Few medical professionals would ever recommend the correct, legitimate cure of exercise (the best thing for high blood pressure) and the use of natural, unprocessed salt. We will begin to explain the benefits of using healthy natural salt instead of the refined table salt in the following chapter.

Chapter Seven—
Clarification: Good/Bad
Healthy Salt vs. Unhealthy Salt

"Refined" Salt is Bad

The real problem with salt is not the salt itself but the condition of the salt we eat – refined! Major producing companies dry their salt in huge kilns with temperatures reaching <u>1200 degrees F,</u> changing the salt's chemical structure, which in turn adversely affects the human body. In the heating process of salt, the element sodium chloride goes off into the air as a gas. What remains is sodium hydroxate which is irritating to the system and does not satisfy the body's hunger and need for sodium chloride. Sodium chloride is one of the 12 daily essential minerals.

Processed Salt- Created Unhealthy

Part of the process for refined salt, or commercial table salt, involves the use of aluminum, ferro cyanide and bleach. To prevent the refined salt from sticking and clumping an anti-caking agents such as sodium ferrocyanide, ammonium citrate and aluminum silicate are added. These are all toxic materials that your body takes in with refined, commercial salt. And because of that process, almost all the vital minerals originally in the real, unrefined salt are removed! One or two servings of refined salt won`t send you to the grave. But continued almost daily use will avail you to the perils of aluminum toxicity. Ferro cyanide is listed by the EPA as a toxic material for human consumption. You are probably aware of the hazards to human health of chlorine, which is used to bleach the salt.[cxxxiii]

Table salt (refined salt) is an inorganic substance that is simply sodium and chloride. It contains no trace minerals and no enzymes. Our bodies struggle to use this enzymatically inactive material without a great cost to our health. When the table salt has been heavily processed, it has become an inorganic chemical. This processed mineral has been rendered nutritionally inorganic. It is very *difficult for our body to break apart* the tightly held together inert mineral of table salt.

Refined salt with*out* the 80 plus elements like magnesium and potassium hinders your cells ability to function properly. Biologically, our trillions of tiny cells are constantly pushing out waste and pulling in nutrients. Cells need a balance of potassium and sodium, along with magnesium, chloride and other trace elements to keep the ionic action working on each side of the cell walls.

Potassium is the principal positively charged ion (cation) in the fluid *inside* of cells, while sodium is the principal cation in the fluid *outside* of cells. Potassium concentrations are about 30 times higher inside than outside cells, while sodium concentrations are more than ten times lower inside than outside cells. The concentration differences between potassium and sodium across "cell membranes" create an electrochemical gradient known as the membrane potential. A cell's membrane potential is maintained by ion pumps in the cell membrane, especially the sodium, potassium-ATPase pumps. These pumps use ATP (adenosine triphosphate) to pump sodium out of the cell in exchange for potassium. Their activity has been estimated to account for 20%-40% of the resting energy expenditure in a typical adult. The large proportion of energy dedicated to maintaining sodium/potassium concentration gradients emphasizes the importance of this function in sustaining life. Tight control of cell membrane potential is critical for nerve impulse transmission, muscle contraction, and heart function.[cxxxiv]

Each cell has this sodium/potassium pump. These charged ions are two of the body's major electrolytes, making the cell "electrically polarized" so that the waste moves out and stays out and the nutrients move in and stay in; this is what is needed for our cells to survive and thrive.

Your cells are very complicated and much beyond the scope of this book. However, it is important to understand the significance of the difference in what we consume and how it affects our bodies. Please take a look at the following sites for more information on how the cell walls work with electrolytes; http://en.wikipedia.org/wiki/Electrochemicalgradient http://science.howstuffworks.com/life/human-biology/nerve4.htm http://www.biochem.cinvestav.mx/PDFs/AlbertsCap11.pdf http://en.wikipedia.org/wiki/Resting_potential Here is a wonderful explanation with a video.[cxxxv]

Processed Salt Epidemic

Salt used to be considered precious. The history of salt is sprinkled with piracy, war, economics, religion, and health. In fact, the next time you contemplate your current salary, consider that the very word "salary" is derived from the Latin word "sal" because Romans often received their pay in salt. If this is hard to accept, consider that during the Age of Discovery, Africans and European explorers traded an ounce of salt for an ounce of gold -- even-steven. Around 110 BC, salt trade was so valued that salt piracy was punishable by death. And Mahatma Gandhi even used salt as major leverage against the British Empire in 1930 when he led thousands of people to the sea to collect their own salt in order to avoid the salt tax imposed by the British.[cxxxvi]

In recent years, salt by being refined has turned into an unhealthy substance that is difficult for the body to process and really should not to be ingested. The refined salt that manufacturers process food with and the salt we use on our tables is so over processed that the molecular structure devitalizes it. It is no longer healthy. Sulfuric acid and chlorine are used to remove the minerals from the natural salt and then a high pressure heat is used to evaporate the water from the salt.

Our consumption of refined, processed salt is one of the biggest health problems we face in our lives today. Salt (refined) is being used for everything in our food supply. This large quantity of toxic chemical is creating havoc in our bodies. Salt is the most element for transportation of fluids in our bodies and this unnatural processed salt is causing an imbalance everywhere in our bodies, and ruining our health.

When we have this deluge of too much refined salt without other minerals to balance it out, water rushes from inside of the cells to the outside through the process of osmosis. The cell then becomes dehydrated! Chronic diseases are created such as cancer. This unbalance caused by *refined* salt can lead eventually to death.

Natural salt with its entire plethora of minerals in our intra and extra-cellular fluid does not move out of the cells and dehydrate them. We need natural, unrefined salt daily. According to an article in "Scientific Integrity for Optimal Health" published by the T. Colin Foundation; "Refined salt does not perform the same functions as unprocessed sea salt, and does not facilitate movement of nutrients into the cells, and removal of wastes from the cells." [cxxxvii]

Table salt is an acidic, inorganic substance

This means the body cannot break the chemical bonds to utilize the sodium and it is not easily excreted from the body. This leads to various health issues, is responsible for 100,000 deaths each year, and contributes to the "sodium is bad" mentality. [Yes, organic salt is alkaline]

Fish from the ocean will die quickly if placed in a solution of refined salt and water. The sodium chloride, in its form (table salt) as it comes from the refinery, is actually in an inert state that is physically not usable to them.

Bottom line is; that yes it can be harmful to consume too much *refined salt*, but you can consume an incredibly elevated amount of natural unrefined salt without deleterious effects. [cxxxviii]

In May 2011, European researchers publishing in the *Journal of the American Medical Association* [cxxxix] reported that the less sodium that study subjects excreted in their urine [an excellent measure of prior consumption] the *greater* their risk was of dying from **heart disease**. These findings call into question the common wisdom that excess salt is bad for you, but the evidence linking salt to heart disease has always been tenuous.

"Lite salt" and "Salt Substitute"

Many well meaning doctors, health providers and medical printed matter have claimed that using a salt substitute could safely cut back on your salt intake. How is that? This is a dangerous medical decision. Lite salt simply uses half salt substitute and half sodium chloride. *Salt Substitues* contains *potassium chloride,* effectively cutting out the sodium content.

But is this the right form of potassium to be ingesting? There are no health benefits to your body using this chemical. You don't have to swallow a whole lot of potassium chloride, as a salt substitute or otherwise, to have big-time problems. Potassium Chloride used for salt substitute, (potassium-40), is mildly radioactive. Huh? Yes, actually all potassium is radioactive and is the reason our bodies show radioactivity with a Geiger counter. Potassium is a strong electrolyte and too much potassium chloride will stop the heart. Curiously, this very same compound has been used on several occasions by Dr. Jack Kevorkian for euthanasia and also in executions. It works by stopping the heart.

"Natural" (unrefined) Salt is Good

Unrefined sea salt contain 98.0 % NaCl (sodium-chloride) and up to 2.0% other micro nutrients called minerals (salts) : Epsom salts, Magnesium salts, Calcium salts, Potassium (Kalium) salts, Manganese salts, Phosphorus salts, Iodine salts and more.

Organic, unrefined salt has at least 80 trace minerals and small amounts of enzymes. This organic chemical (natural salt) with its carbon, enzymes and minerals is used eagerly by your body. This nutritionally organic salt is chelated (bound) to other organic molecules in our bodies. The organic, unrefined salt is easily broken apart and used.

All together, over 100 minerals composed of 80 chemical elements... Composition of crystal of ocean salt is so complicated

that no laboratory in the world can produce it from its basic 80 chemical elements. For this reason *all testing* and health predictions are done with pure NaCl which does not replicate the ocean of sea salt running through your veins.
Nature is still better chemist than people.

Many people believe that salt is harmful to the human body. The truth is we cannot live without salt (sodium chloride). From salt the body makes hydrochloric acid which is one of the essential digestive fluids. Mostly there is not enough natural salt in our foods we eat, so we must supplement our diet. When salt is withheld, weakness and sickness follow.[cxl]

Salt Experiment—Natural vs. Refined

Try this experiment: Mix a spoonful of salt in a glass of water and let it stand overnight. If the salt collects on the bottom of the glass, it has been processed. NATURAL SALT DISSOLVES! Salt that will not dissolve in water cannot dissolve in your body.

Any foreign substance that collects in the body organs and tissues will eventually result in malfunctioning of essential body processes: heart disease, arthritis, hardening of the body tissues and arteries, calcium deposits in the joints, etc. Natural organic salt (saline) is accepted in the body and will not cause calcification and can actually help dissolve damaging calcium deposits in the body.

Processed foods manufactured and set on shelves for us to purchase use a lot of salt. This increases it shelf life and improves its taste. The salt used in much of this processed foods is not from raw salt, but rather is a chemical salt created in a laboratory.

Whether chemically created or collected and over processed, it is far different from *natural* sea and mineral salts which can actually be beneficial to health. It is important to recognize the difference between common table salt, which is not natural, and natural earth salts when making any recommendations pertaining to salt. [cxli]

Many studies have been conducted linking sodium to many of American's modern day health problems; however, we need to point out the fact, these studies are conducted using table salt (sodium chloride) or isolated sodium (sodium that has been stripped of the minerals needed for the body to utilize it properly). The table salt found in high amounts in the standard American diet is not synonymous with the "un-tested" good type of sodium our body needs to maintain good health.

The Virtues of Naturally Organic Salt—

According to, Dr David Brownstein, (author, "Salt Your Way to Health"), unrefined salt is an excellent detoxification aid, as well as a provider of mineral nutrients in a naturally bio-available balance. There are usually around 80 minerals and essential trace elements in unrefined, organic salt. Soil grown food is lacking in many of these because the soil has been depleted of trace elements and minerals. These minerals are, Magnesium, calcium, potassium, and sulfate. Obviously, sodium is present also, but it comprises only 50% of the total mineral content rather than the 99% sodium in refined table salt. [cxlii]

True organic sodium is essential, beneficial, and needed by our body in moderate amounts. This type of sodium is found in fruits, vegetables, natural sea salt and soil. Sodium is known as the "youth mineral" because it is associated with youthful, limber, and flexible joints. The alkalinity of organic sodium helps neutralize acids that result from stressful lifestyles and poor nutrition. Without adequate amounts of sodium, the body will take minerals from its reserves, including the bones, to neutralize acid. Organic sodium is essential for calcium absorption, digestion, bile production, fluid balance, and the function of the brain, kidneys, liver, lymph, blood, spleen, gastric secretions, cellular function, and metabolism. Unlike table salt, an excess of this type of sodium is easily excreted from the body.[cxliii]

Sea salt is highly alkalizing

Sea salt is naturally very alkalizing for the body, while iodized table salt is highly acid forming. Table salt is pure sodium chloride (with aluminum added for anti-caking) and chemically neutral due the chemical reaction of the base sodium and chloride combining.

Salt as it is found in nature is very healthy. One of the reasons is because sea salt contains a wealth of trace minerals that are essential for good health. The problem, is that common table salt no longer contains these trace minerals. The "food police" and other advocates of salt-free or low-salt diets are ignoring this important point. It is not salt that is bad; it's the contaminated and over processed salt that most people use.

Refined salt, stripped of its natural mineral structure, is virtually all sodium. As a matter of fact, it is up to 99% sodium! That's why it is called Sodium Chloride. Processed foods are laced with Sodium Chloride refined salt as well as with other types of sodium, such as sodium benzoate, sodium nitrate, and the notorious monosodium glutamate (MSG). So if you are concerned about your sodium levels, then cut out processed foods!

Use only sea salt, not the white (bleached out) stuff. A good sea salt is grey looking. The "food police" lump all salts together as bad for you.

Recognize Pure Unrefined Salt

Unrefined salt has a distinctively different look from refined salt. It is usually too course to be used in a salt shaker. You may want to invest in a salt grinder. And it is usually not very white. Off white is more common, even pink or gray for unrefined pure salt. The extreme white of common household or commercial salt is a result of bleaching. But buyer beware, some so called sea salts offered on line and in health food stores are at least partially processed. Avoid sea salt that is too white and too fine as a rule of thumb.

If you are very concerned about getting the purest available product, and you don`t have anyone`s advice you can trust, look for "organic certification". Since salt is mined or taken from salt water beds, organic has different implications than produce and animal product organic requirements. But the standards are there and they are strict. The two groups that certify salt as organic are BIO-GRO of New Zealand, and Nature & Progresre of France.

Organic Salt

Organically produced, unrefined salt should be a healthy addition to our diets. It offers bio-available, balanced minerals that aren`t naturally present in our food chain. It does not contain the poisons of industry that are a part of refined salt. Yes, too much of a good thing can be bad. But again, the sodium of refined salt and other food additives is curbed best by --eliminating processed foods--, which contain several toxic sodium sources as well as unrefined salt, from the diet. Using organic, unrefined salt seems like a natural and economical way to boost one`s immune system.[cxliv]

Table salt is mined from underground salt mines or sometimes from open salt beds. It is then heavily processed, refined, stripped of all its minerals except for sodium and chloride (NaCl), and subjected to anti-caking agents like aluminum.

Organic salt is alkaline

Table salt (NaCl) and sodium "sea salt" are not synonymous. In fact, the majority of Americans are deficient in the good type of sodium the body needs. True organic sodium is essential, beneficial, and needed by our body in moderate amounts. This type of sodium is found in fruits, vegetables, natural sea salt and soil.

Sodium is known as the "youth mineral" because it is associated with youthful, limber, and flexible joints. The alkalinity of organic sodium helps neutralize acids that result from stressful lifestyles and poor nutrition. Without adequate amounts of sodium, the body will take minerals from its reserves, including the bones, to neutralize acid. Organic sodium is essential for calcium absorption, digestion, bile production, fluid balance, and the function of the brain, kidneys, liver, lymph, blood, spleen, gastric secretions, cellular function, and metabolism. Unlike table salt, an excess of this type of sodium is easily excreted from the body.[cxlv]

Salt from Plants!

True organic <u>sodium</u> is essential, beneficial, and needed by our body in moderate amounts. This type of sodium is found in fruits, vegetables, natural sea salt and soil. To prevent a deficiency of this important mineral, ditch the table salt and add sodium rich foods to your diet. Foods highest in organic sodium include:

- celery
- asparagus
- barley
- red cabbage
- carrots
- coconut
- okra
- lentils
- kale
- strawberries
- sesame seeds
- raisin
- goat's milk
- egg yolks
- pure (non-iodized) sea salt.

Sodium is known as the "youth mineral" because it is associated with youthful, limber, and flexible joints. The alkalinity of organic sodium helps neutralize acids that result from stressful lifestyles and poor nutrition. Without adequate amounts of sodium, the body will take minerals from its reserves, including the bones, to neutralize acid. Organic sodium is essential for calcium absorption, digestion, bile production, fluid balance, and the function of the brain, kidneys, liver, lymph, blood, spleen, gastric secretions, cellular function, and metabolism. Unlike <u>table</u>

salt, an excess of this type of sodium is easily excreted from the body.[cxlvi]

Deficiency of good salt

We have discussed this in detail earlier in the book but I want to mention this as it relates to our "good salt" theme here. There are many health problems caused by an organic sodium *deficiency* including: 1. gallstones, 2. kidney stones, 3. hardened arteries, 4. osteoporosis, 5. arthritis, 6. gout, brittle bones, 7. heartburn, 8. acid reflux, 9. gastroparesis, 10. nerve problems, 11. indigestion, 12. headaches, 13. stiff or painful joints, 14. abnormal pulmonary function, 15. bacterial infections, 16. poor memory, 17. diabetes, 18. bloating, 19. fatigue, 20. restless legs, 21. weight gain, and 22. headaches, and more.

Organic sodium from *food* is the best source of salt and is not responsible for poor health, as celery or any of the other foods listed above have not been used as controls in the research and studies responsible for the negative sodium publicity. The vast amount of health issues attributed to "sodium" are directly related to the high amounts of table salt (refined) found in the processed foods most Americans consume. To prevent these health issues, ditch the table salt, not the sodium.[cxlvii]

"Preppers"-survivalists

Most people vastly underestimate the amount of salt they will need in a survival scenario. This happens because all our food is pre-salted right now by food manufacturers. So we don't think we need extra salt. But in the absence of processed, pre-salted food, an individual's need for real (unrefined) salt is surprisingly high.[cxlviii]

Salt is essential for life.

That's why every land animal seeks out salt and craves it. It's why ocean creatures *live* in waters made out of salt. It's why human blood is fundamentally made with salt. (blood is .9 salinity) Your body cannot retain water without salt, and when a hospital drips an IV solution into your bloodstream to keep you alive, it's a salt (saline) solution!

Your body was engineered to *like* the taste of salt because your body absolutely needs salt's minerals in order to survive.[cxlix]

For decades Americans have been told by food and diet experts in government and the private sector that salt, more than any other substance, is harmful to our health, even more than fats, alcohol, sugars and, presumably, bad seafood. Now, there is new evidence that the contents of your table salt shaker are not as bad as once believed, even without the benefits of the healthy natural salt we have been touting.

Salt Your Food

Researchers from McMaster University in Ontario looked at data from drug trials involving nearly 30,000 individuals who already had heart disease or diabetes. Participants in these trials had their sodium intake measured through urine analysis and were followed for an average of four to five years to record the incidence of heart-related hospitalizations and deaths.

After adjusting for factors like medications, weight, smoking and cholesterol levels, researchers found that too little salt is doing harm instead of good. Those who consumed between 4,000 and 6,000 milligrams of sodium per day--more than double the current recommendations--were at the *least* risk for heart disease and stroke. Read again----they had <u>less risk</u> of heart disease with <u>more salt</u>![cl]

People who ate a diet lower in salt didn't experience less risk, but more! Researchers found that people who consume 2,000 to 3,000 mg of sodium per day were actually 20 percent more likely to experience death or hospitalization related to heart conditions, compared to those consuming between 4,000 and 6,000 mg daily.[cli]

Moderation-- Common sense

But don't take this as advice that salt intake should be completely unlimited. There are reports that thousands of people are eating rock salt for health benefits sold from health food stores. It is reported to be an excellent way to assimilate minerals that are in natural salt.[clii] [Makes me pucker thinking about it] Moderation in everything we consume is important and that does include consuming too much salt. ~~~Should I offer "Rock Salt" at our next dinner party?

Natural Salt

Natural salt like <u>sea salt</u>, of course is the healthy, safe answer. Never use highly refined commercial salt, which often contains harmful additives and lacks a balanced mineral profile.
Dr. Martin O'Donnell, lead author of the study and associate clinical professor of medicine at McMaster University, says, "When you take people at more moderate intake levels, there is emerging uncertainty as to whether there are long-term benefits of reducing sodium intake further."

The new report, published in this week's issue of the Journal of the American Medical Association, contradicts what many of us have been told about salt. The research team involved urges officials to recommend a safer range of sodium intake rather than to set a single rigid limit. Results from this study indicate that people who already consume a moderate amount of sodium do not benefit from lowering their salt intake. In fact, it may even harm them.[cliii]

Erroneous reports on Sea Salt

You have seen many reports lately that state that unrefined sea salt has no added benefits over the white processed table salt. Not so fast. These reports fling the words "scientists", "experts" or "doctors" and claim that sea salt has no health benefits. I do not know what agenda would drive these statements, but there have been no scholarly documentation from a reputable lab to show this.

One example is a bogus claim made denying the benefits of sea salt was made recently by CASH. According to a report released by pressure group Consensus Action on Salt & Health (CASH), unrefined and 'gourmet' salts containing trace and ultratrace minerals are no healthier than refined table salt.

Following is a statement of rebuttal by "The Alliance of Health".

"I am confused. Firstly, surely more expensive salt encourages people to use less? Secondly, [CASH Chairman] Prof MacGregor has arrived at sweeping conclusions claiming the salt has no health benefits when they've done no research to test whether it does or does not. Others have. The best example is the "Pansalt case" [cliv] from Finland, where a dramatic reduction in hypertension rates has occurred over 30 years, not just because of reduced sodium, but because of the addition of other minerals included. And then there's years of clinical experience. Naturopaths over decades have found profound effects in using unprocessed sea salts as compared with synthetic sodium chloride, which is missing the trace elements. Three percent other ingredients in the Himalayan [salt] example Professor MacGregor [cited] may not sound a lot, but, tiny amounts of trace and ultra[trace] elements are needed for enzymatic and other systems in the body. Hydroponics and intensive agriculture means our food is greatly depleted in these minerals. Professor MacGregor should do some research on unprocessed sea salts and check out the results for himself."[clv]

Fluoride

You may come across an analysis of organic salt minerals, or a commentary on such, which mentions <u>fluoride</u> as a constituent. But there are two types of fluoride. One, Calcium Fluoride, is an element that occurs as a natural process over time within the earth's soil, rock, and water areas. This is the fluoride that *originally* was claimed as a deterrent against tooth decay. *Wikipedia* notes that while all other fluorides are dangerous for human consumption, calcium fluoride is not. And it's Calcium Fluoride that would be in any unrefined *natural* salt analysis.

Calcium Fluoride is okay.

Sodium Fluoride is hazardous. [clvi]

Sodium Fluoride is purchased from China (who lists it as an "insecticide") and added to America's tooth paste and city water supplies. Yes, most US city's water we drink and cook and bath in has a very poisonous chemical called Sodium Fluoride that destroys the pineal gland and your thyroid![clvii]

Ever notice how many people are on thyroid medicine? Well, you should. There are 27M people with Thyroid disorders here in the United States. Thyroid cancer is on the rise, with 56.5M new cases in 2012 alone! It has doubled since 1990.This is an epidemic! Sodium Fluoride is a major contributor to Thyroid disease. [Infections and too many x-rays are other causes.] Due to the lack of screening, actually, 60M Americans could have poorly acting Thyroids! And we bath in Sodium Fluoride, we drink it in our city water and we even brush our teeth with it. [clviii]

This synthetic, poisonous fluoride is a waste by-product of the aluminum industry, fertilizer industry, and nuclear industry. It`s their way of picking up a lot of easy bucks by selling it to municipalities for their water supplies instead of suffering the expense of getting rid of it. [clix]
It causes health problems, and should never be ingested into or onto your body! Have you heard the phrase "dumbing down"? Well, guess what this fluoride does according to a report from China? Yep. A recent Chinese study concluded that low dose sodium fluoride in drinking water diminishes IQ, especially among children.
So it is one more source of "dumbing down" of America. Either the dentist's do not distinguish between those two, or perhaps they do not even know there are two types of fluoride.

BTW: Sodium Fluoride also blocks iodine from being absorbed into your body. Jumps in first and fills the ionic spot so you get no iodine. Read this and watch a real case video of what fluoride has done to real people's health;
heiodineproject.webs.com/fluoridegoodorbad.htm

Don't brush with sodium fluoride! [clx] Go to a health food store and pick up an all natural tooth paste and you will be surprised to find you will not miss the old bad stuff. I use the organic toothpaste *Jason* peppermint and it tastes delicious.

Chapter Eight—Salt, We Misjudged You

Myths

Two of the biggest health myths of today are about salt. These myths are that salt is bad for you and that it causes hypertension. Neither is true. Only refined/processed salt is bad.
Salt *unrefined and in a natural form* has a wide variety of health giving minerals; including magnesium, potassium and calcium. According to the National Health and Nutritional Examination Survey in 1984, low levels of *minerals* were associated with hypertension. Low mineral intakes are associated with elevated blood pressure. Low salt diets are dangerous. *Refined/highly processed* salt only contains sodium and chloride!

The potassium from the *natural* salt moves into the cells and draws extra-cellular fluid *out*! This lowers the blood volume outside the cells therefore the blood pressure. Only Natural Salt gives this benefit.

Processed Salt Creates a Deficiency of Salt!

> ➤ A low-salt diet causes higher total cholesterol and lousy LDL cholesterol levels. [clxi]
> ➤ A low-salt diet increases hormones aldosterone, rennin, angiotension and noradrenaline—which cause heart attacks.

- A low-salt diet raises insulin levels, bringing on diabetes, obesity and other bad health problems.[clxii]
- A low-salt diet allows the accumulation of toxic elements in the body, leading to delirium, psychomotor retardation, schizophrenia and aging.
- A low-salt diet creates a deficiency in minerals. Unrefined salt provides the balance of minerals to ensure the proper thyroid functions.[clxiii]

MYTHS ABOUT SALT

#1: We eat more salt today than ever before.

FACT: Our current salt consumption (1.5 to 1.75 teaspoons per day, 8-9 grams) is about one half of the amount consumed between the War of 1812 and the end of World War II, which was about three to 3.3 teaspoons (16-17 grams) of salt per day.[clxiv]

#2: Our *knowledge* of the major sources of salt in our diet (i.e., 80 percent from processed foods) is unquestionable.

FACT: These data, referred to in every medical publication, is based on a single paper from 1991, which involved a dietary recall (a very unreliable method of data gathering) of a total of just sixty-two persons.[clxv]

#3: Our salt consumption continues to rise every year.

FACT: There has been no change in our consumption of salt since the mid-1950s.[clxvi]

#4: The thirty-year public health initiative in Finland represents a successful model of salt reduction.

FACT: While Finland was able to reduce salt consumption among its population from 2.3 teaspoons of salt per day down to 1.3 teaspoons from 1970 to 2000, the health benefits that they have achieved during the same time period were no better (marginally worse) than neighboring and other countries that did not reduce salt consumption.[clxvii]

#5: Current levels of salt consumption result in premature cardiovascular disease and death.

FACT: When average life expectancy in various countries is plotted against the average salt intake in those countries, it is clear that the higher the salt consumption, the longer the life expectancy. While no cause-and-effect relationship between sodium intake and lifespan is implied, the data clearly demonstrate the compatibility between life expectancy and the associated levels of sodium intake.--InterSalt Life Expectancy[clxviii]

#6: Cutting back on salt will improve the overall diet.

FACT: Salt makes the bitter phytochemicals in salad greens and vegetables more palatable. Removing salt from dressings or accompaniments will make these important diet items less acceptable and will discourage people from eating them.[clxix]

#7: Reduced salt levels are critical to the DASH diet.

FACT: When the results of the DASH Sodium trial are examined it is immediately apparent that merely moving to a DASH diet (red line) has a significantly greater impact on blood pressure than simply lowering salt consumption. Dropping from the normal level of sodium consumption to the Dietary Guidelines' recommended level reduced the systolic pressure in the American diet by a mere average of 2.1 mm Hg. However, simply changing from without any changes to sodium consumption reduced the systolic blood pressure by 5.9 mm Hg. Almost three times the drop resulting from sodium reducing! More important, reducing salt makes the DASH diet far less palatable and thus discourages people from adopting it.[clxx]

#8: There is a clear relationship between salt intake and blood pressure.

FACT: The lack of a clear relationship between salt intake and blood pressure is best exemplified with the standard hospital saline IV drip, which supplies an average of three liters of 0.9 percent sodium chloride per day. This is equivalent to twenty-seven grams of salt (4.5 teaspoons) per day while in the hospital in addition to the six grams (one teaspoon) of salt taken in food (if the Guidelines are followed). That is a total of thirty-three grams of salt per day or more than five times the Dietary Guideline recommendations! Yet patients' BP is checked every four to six hours and does not change.[clxxi] Different science?

#9: Reducing salt intake can do no harm.

FACT: Reduced salt intakes have repeatedly been linked in the medical literature to the following conditions: Insulin resistance (diabetes), metabolic syndrome, Increased cardiovascular mortality and readmissions, cognition loss in neonates and older adults, Unsteadiness, falls, fractures, Lifelong avidity for salt, and more[clxxii]

#10: The U.S. Dietary Guidelines process is valid.

FACT: The original Dietary Recommended Intakes (DRI), issued under the imprimatur of the Institute of Medicine (IOM) (National Academy of Sciences), were immediately accepted internationally and spared the critical scientific review normally given to nutritional recommendations.

Several participants stressed that the DRIs were largely based on the lowest quality of information—opinion—rather than on randomized controlled clinical trials which represent the highest quality of evidence.

The 2005 DGA for sodium referred to the DRI's as a foundation document and assumed all its recommendations. In 2010, DGAs reconfirmed the recommendations of the 2005 DGAs with the provision that the at-risk populations consume 1,500 mg sodium 1/2 teaspoon per day for the upper limit.

Then, as it happened, the Chair of the original DRIs committee that set the first recommendations for sodium also happened to serve as the Chair of the 2005 Dietary Guidelines Subcommittee on Electrolytes and thus evaluated the very recommendations that

the chairperson was responsible for promulgating in the first place.[clxxiii]

Chapter Nine—Toxins and Sweat

Since sweating involves salt, let's clear the air about some misconceptions regarding this secretion and have a talk about "sweat".

First of all, in contrast to the vast quantities of literature dedicated to the analysis of blood and urine, the secretions of the sweat gland have been investigated relatively poorly and, therefore, are rarely used in clinical practice. Therefore, this lends itself to mistaken beliefs about our friend, "sweat".

We have been as-salted! Pick up a magazine while sitting in the waiting room or cruse through a newspaper like the Los Angeles Times and you might see the latest article on health, telling you that *toxins are not removed with sweat.* It is such an "in-thing" to present this wonderful information. Someone writes an article somewhere and all the publishers in the land are looking for print material and they have hit this information to the public with a lucrative zeal.

Look, I "get it", that the purpose of sweating is to cool down the body. Furthermore we know that the lymph system, kidney and liver are the bastions of toxin removal. None of which, the sweat glands are directly connected to.

Before we go through our contraire opinion about the substance called "sweat", we will run through the science of **toxins**.

One of the body's regulatory apparatuses is its ability to store toxins and unusable materials in the comparatively inactive tissues--bone, cartilage, connective tissues--pending its elimination at a more favorable opportunity. However, the favorite site for such deposits is the *subcutaneous* connective tissue. [clxxiv]

Our skin [called hypodermis or subcutaneous] is the largest organ of the body and contains a nice layer of fat with a considerable amount of toxins. Sweat glands reach down lower than the hair follicles and throughout the fatty layer. Yet, the media insists mom was wrong and that when our skins leak from the heat, it is only water and a little salt.

It is hard to find any dependable, scholarly science experiments that are well documented with published results to show what is truly in sweat. How disconcerting. I feel let down. Two points can be made here:

Common sense tells us that if the subcutaneous tissue was rampant with toxins the sweat gland should be infiltrated with the liquids surrounding and sent out the porthole (skin pore). Also...If the body was replenished with healthier water and salt, perhaps we have diluted the bad.

There are other scientists that think toxins come out with sweat and some might surprise you.

1. Forensics. *Read article from Forensics Science International.*[clxxv]
2. Law Enforcement for Cannabis detection.[clxxvi]
3. Private professional laboratories.[clxxvii]

Science of Sweat

Sweat is a dilute electrolyte solution excreted by the eccrine (sweat) glands in the skin. Although the primary function of sweating is to control body temperature, it is loaded with other things like minerals, toxins and fat. The male sweat also serves as **chemosignals** that influence the hormonal balance of females, and therefore act as pheromonal stimuli.

Unbelievable! Why…do the popular health information providers such as magazines, health pamphlets, Google or almost everywhere on the internet, claim that sweat is just water and a little NaCl?

One reason for the lack of knowledge in medicine and biology is explained by the difficulties of *sweat sampling* in sufficient quantities for analytical work. Although the composition of sweat is primarily water, studies have shown that various organic and inorganic compounds are also present.
They claim sweat does not contain other substances and toxins. Hogwash!
It does. Sweat has chemicals and wastes that come out of your body. Though hundreds of search engine sites tell you sweat is water and salt, there are honest hard working scientists publishing data from their research that show results of their "sweat findings". Let's take a quick look at real science.

Peer reviewed science articles

1. MD'3 ABSTRACT by; M Brune, MD, B Magnusson, MD, H Persson, and L Hallberg

The losses of iron in whole body cell-free sweat were determined in eleven healthy men. The findings indicate that iron is a constituent of sweat. [clxxviii]

2. *Analyzing and Mapping Sweat Metabolomics by High-Resolution NMR Spectroscopy* by; Viktor P. Kutyshenko mail, Maxim Molchanov, Peter Beskaravayny, Vladimir N. Uversky mail, and Maria A. Timchenko. [clxxix]

There is an almost steady level of free amino acids in the sweat samples. Free amino acids accumulated in the epidermis are not reabsorbed and are excreted with the sweat gland secretions.[clxxx]

According to their findings this team has reported on this research article that sweat contains lactate, Glycerol and the distinctive component of sweat... pyruvate. They report: *"Since the sweat and sebaceous glands are parts of the excretory system, they excrete the **undesirable** products from the organism into the environment, e.g. urea, ammonia, **prescription drugs**!*

Yes, prescription drugs are eliminated from your body through sweat. Drugs like: griseofulvin, ketoconazole, etc.

*Some healthy people have rather high levels of **sucrose** in their sweat."*
The sweat from the lower regions of the back showed more secretion of **fats** and greater content of **triglycerides.** The feet skin contained large quantities of **serine** and **urocanic acid**. [clxxxi]

The partial list of known substances in sweat:
- Amino acids
- Lactate
- Glycerol
- Pyruvate
- Urea
- Ammonia
- Prescription drugs
- Sucrose
- Fats
- Triglycerides
- Serine

- Urocanic acid
- Antimicrobial peptide (regulation of human skin flora)[clxxxii]

Sweat glands

Sweat glands are simple tubules with a duct that carries the sweat away. The base of the sweat gland is set deep in the hypodermis, and the entire gland is surrounded by adipose tissue. All sweat glands' secretory coils are wrapped in long, rod-like contractile cells. Excretory ducts carry sweat away from the secretory coil.

Apocrine gland secretions may also contain pheromones, chemicals that communicate to other individuals by altering their hormonal balance. Some research has indicated that feminine secretions from apocrine sweat glands can alter the menstrual timing of other women. The significance of human pheromones and the role of apocrine *sweat gland secretions* in humans *remains incompletely understood.*[clxxxiii] [A quote, not my statement]

Eccrine sweat glands are smaller and found on soles, palms, and scalp.

The clear secretion produced by eccrine sweat glands is termed sweat or *sensible perspiration*. Sweat is mostly water, but it does contain some electrolytes, since it is derived from blood plasma, although less concentrated. It therefore contains mainly sodium chloride, but also other electrolytes. The presence of sodium chloride in sweat gives it a salty taste.

The total volume of sweat produced depends on the number of functional glands and the size of the surface opening. The degree of secretory activity is regulated by neural and hormonal mechanisms (men produce greater volumes of sweat than women). When all of the eccrine sweat glands are working at maximum, the rate of perspiration for a human being may exceed three liters per hour, and dangerous amounts of fluid and electrolyte losses can occur.[clxxxiv]

---Three liters per hour? That would be about ten percent of your whole body's liquid. Yikes! No wonder I stop working out in twenty minutes.

Functions of Sweat

Besides the effective accumulation of various excretory products, sweat serves important biological jobs for the body.

The glomeruli and ducts of the eccrine sweat glands are known to intensely synthesize various proteins, including the large quantities of bactericidal and regulatory peptides. These peptides restrict the development of pathogenic and symbiotic micro flora, and determine the normal functionality of the enzymatic and excretory systems of skin cells. [clxxxv]

Sweat assists the immune system when the hypothalamus increases body temperature and kills bacteria, infections and even viral infections.

The sweating process actually burns calories.

"Night Sweats"

Most people experience occasional incidences where you wake up drenched with sweat. *What is happening?*—is a natural response. When the shock wears off, relax. Your body is working on your health. Raising the body temperature, causing it to sweat, kills bacteria and infections. Through the sweat glands, the body is able to "dump" unwanted material. Following is a list of medical conditions known to cause the body to pull out this "tool" in its arsenal of health methods.

- Alcohol. Especially if you have had "one too many" triggers sweat to detoxify.[clxxxvi]
- Low blood glucose or insulin. [clxxxvii]
- Medications. Aspirin, acetaminophen.[clxxxviii]
- Medications. Prescription drugs. [clxxxix]
- Infections. Can be low grade you are not aware of. Especially tuberculosis.[cxc]
- Cancer. Especially lymphoma.[cxci]
- Hyper thyroidism. [cxcii]
- Hormones.[cxciii]
- Menopause. [cxciv]
- Brain excited from dreams.
- REM, the deep sleep mode.
- Lymph system. Body clearing out the lymph system. (collects toxins and infections)[cxcv]
- Immune system. Assists the immune system.[cxcvi]
- High metabolism.
- Recent activity. Often triggers night sweats.

Summary

I will leave this damp subject matter for your further investigation. Please go to the following site and scroll down to the bottom to see a list of thirty two (32) great scientific reports from laboratories:

http://www.plosone.org/article/info%3Adoi%2F10.1371%2Fjournal.pone.0028824

These reports all show the results of toxins and chemicals secreted through the sweat glands. These sources can be hard to read due to the scholarly scientific jargon and reporting in chemical terms. Easier reading can be found through simpler sites like http://www.3fatchicks.com/the-benefits-of-sweating/; however, they do not give the professional published scientific articles.

Still more studies are needed for the science if how sweating effects the health of the body. There is no scientific consensus as of yet regarding optimal salt intake, since body sodium levels also depend on how much an individual sweats.

Chapter Ten—Water Retention

Since salt is automatically blamed for bloating and water retention, we will briefly discuss bloating or water retention. First, 70% of the population is not salt sensitive and would not be affected by salt/sodium and no water retention would occur. In the remaining 30% of people with sensitivity to salt, we need to understand that there is a whole plethora of chemical reactions and biological process going on in our body that creates bloating.

The term <u>water retention</u> (also known as fluid retention) signifies an abnormal accumulation of fluid in the circulatory system or within the tissues or cavities of the body.

Water is found both inside and outside the body's cells. It forms part of the blood, helping to carry the blood cells around the body and keeping oxygen and important nutrients in solution so that they can be taken up by tissues such as glands, bone and muscle. Even the organs and muscles are mostly water.

The body uses a complex system of hormones and hormone-like substances (called prostaglandins) to keep its volume of fluid at a constant level. If one were to intake an excessive amount of fluids in one day, their weight would not be affected in the long-term because the kidneys quickly excrete the excess in the form of urine. Likewise, if they do not get enough to drink, their body will hold on to its fluids and they will urinate less than usual.

The tissues of the body are adapted to a specific osmotic pressure and as soon as this pressure is exceeded, the substance responsible for the excess is automatically excreted by the kidneys. When this rise in osmotic pressure is due to _NaCl_ the process of excretion has an inhibiting effect upon kidney function.

Yes, both sodium and chloride hinder the normal excretion of water by the kidney cells. Salt retards kidney function, and at the same time raises the osmotic pressure throughout the body. However, with the right kind of salt and minerals and vitamins along with some effort to properly hydrate with water, the kidney and body should function very well. All excesses are harmful and in fairness, salt taken in excess over long periods of time can affect the lining of the endothelial lining of the blood vessels.

Body water retention, edema, high blood pressure and swelling causes' pain, discomfort and creates difficulty even moving around. This is a bad thing, however in our society we have been quick to blame one thing and have created a solution—delete the "bad guy". According to the professionals, government and the general everyday public...the culprit is salt! Seldom researching other causes and never considering the form of salt we are ingesting! Millions of Americans are on a salt free or low salt diet "just in case". Let's take a look at inflammation a little further.

Other causes of water retention in the body---

- Leaky capillaries
- lymphatic system
- lack of exercise
- protein
- histamine
- hormones

- Acute kidney failure
- Cardiomyopathy (disease of heart tissue)
- Chronic kidney failure
- Chronic venous insufficiency (problem with leg veins returning blood to the heart)
- Heart failure
- Hormone therapy
- Lymphedema (blockage of lymph system)
- Nephrotic syndrome (damage to small filtering blood vessels in the kidneys)
- Nonsteroidal anti-inflammatory drugs (NSAIDs), such as ibuprofen (Advil, Motrin, others)
- Pericarditis (swelling of the membrane surrounding the heart)
- Preeclampsia (pregnancy-induced high blood pressure)
- Pregnancy
- Prescription medications, including some drugs for depression, diabetes and high blood pressure
- Prolonged sitting, such as during airline flights
- Prolonged standing
- Thrombophlebitis (blood clot, usually in the leg)
- Achilles tendon rupture
- ACL injury (tearing of the anterior cruciate ligament in your knee)
- Baker's cyst
- Broken ankle/broken foot
- Broken leg
- Gout (arthritis related to excess uric acid)
- Infection or wound in the leg
- Knee bursitis (inflammation of fluid-filled sacs in the knee joint)

- Osteoarthritis (disease causing the breakdown of joints)
- Rheumatoid arthritis (inflammatory joint disease)
- Sprained ankle

Swelling, edema is the impairment of the balance between the secretion and removal of fluid in the interstitium. Any condition that increases or reduces the oncotic pressure (low plasma osmolality) will cause edema. Increased hydrostatic pressure will likewise have the same effect. Obstruction of the lymphatic system, a subsystem of the circulatory system, may result to abnormal removal of the interstitial fluid and will also lead to edema.

Age, bloating and weight

As we get older we gain weight. Our whole body swells with interstitial fluid. This "water" hangs around, makes us bigger, though our muscles are getting smaller. This "water" gives us weight gain with no benefits. Diets do not get rid of this! It is harder and harder to lose, to get out of our body due to the lack of properly functioning health. The circulatory system, lymph system, kidney and liver cannot keep up with all of the things listed above eventually. ***Kinda-ticks me off but that is the way it is.

Capillaries are too leaky

Fluid rich with oxygen, vitamins and other nutrients passes all the time from the capillaries (the smallest blood vessels) into the surrounding tissues, where it is known as tissue fluid or interstitial fluid. This fluid nourishes the cells and eventually should return to the capillaries. Water retention is said to occur as a result of changes in the pressure inside the capillaries, or changes that make the capillary walls too leaky. If the pressure is wrong, or the capillaries are too leaky, then too much fluid will be released into the tissue spaces between the cells. Sometimes so much fluid is released that it cannot all return to the capillaries and remains in the tissues, where it causes the swelling and water logging which is experienced as water retention.

Lymphatic system

Another set of vessels known as the lymphatic system acts like an "overflow" and can return a lot of excess fluid back to the bloodstream. But even the lymphatic system can be overwhelmed, and if there is simply too much fluid, or if the lymphatic system is congested, then the fluid will remain in the tissues, causing swellings in legs, ankles, feet, abdomen or any other part of the body.

Lack of exercise

Lack of exercise is another common cause of water retention in the legs. Exercise helps the leg veins work against gravity to return blood to the heart. If blood travels too slowly and starts to pool in the leg veins, the pressure can force too much fluid out of the leg capillaries into the tissue spaces. The capillaries may break, leaving small blood marks under the skin.
The veins themselves can become swollen, painful and distorted - a condition known as varicose veins.

Lack of exercise is a common cause of water retention, because muscle action is needed not only to keep blood flowing through the veins but also to stimulate the lymphatic system to fulfill its "overflow" function. Long-haul flights, lengthy bed-rest, immobility caused by disability and so on, are all potential causes of water retention. Even very small exercises such as rotating ankles and wiggling toes can help to reduce it.

Protein excess

Protein attracts water and plays an important role in water balance. In cases of severe protein deficiency, the blood may not contain enough protein to attract water from the tissue spaces back into the capillaries. This is why pictures of starving people often show an enlarged abdomen. The abdomen is swollen with edema or water retention caused by the lack of protein in their diet.

When the capillary walls are over-permeable (too leaky), protein can leak out of the blood and settle in the tissue spaces. It will then act like a magnet for water, continuously attracting more water from the blood to accumulate in the tissue spaces. [cxcvii]

Hormones

A complex system of hormones and prostaglandins (hormone-like substances) is used by the human body to regulate water levels. So that excess water intake one day can be resolved by the kidneys quickly excreting the excess urine, while a lack of fluids on another day may result in much less urination than usual.

Histamine

When an inflammation is present in the body, *histamine* is released. Histamine causes the gaps between the cells of the capillary walls to widen, making them leak more. The aim is to make it easier for infection-fighting white cells to quickly get to the site of an inflammation (infection). However, if the inflammation persists for a long time, water retention can become chronic (long-term).

Medication

Almost any prescription drug causes swelling. [cxcviii] Read the label, check with the pharmacist or doctor. Try to find some scientific tests published, however, any drug causes a reaction and side effect, which ultimately results in...swelling-edema-retention.

The Waterfall Diet

Since we are on the subject of water retention, I am just mentioning something that millions have tried. The aim of the "Waterfall Diet" is to release water retention through urination. It does not work by stimulating the kidneys. The diet is high in flavonoids and some other nutrients which accelerate the repair of leaky capillaries.

The Waterfall Diet also helps the user identify any foods which the body is not digesting properly, resulting in higher histamine release. Celery and parsley, as well as other coumarin-rich foods play a key role in this diet - coumarin helps macrophages (type of white blood cells) break up proteins which have leaked into the tissue space.

Chapter Eleven—Salt and Hypertension

Most people are not Salt Sensitive

What? Yes, that's right. What does this mean?

One-third of all hypertensive people are salt-sensitive and will benefit from a low-sodium diet. *Most people are not affected by a salt regimen.* Your feet, your heart or your body would not swell with any amount of salt consumption, unless you are in the 30% of confirmed hypertension statistics.
http://www.madehow.com/Volume-2/Salt.html#b#ixzz2OgIs0W8m

Since there is no way to tell who these people are, most hypertensive persons under medical care will be placed on such a diet to see if it helps. While some have suggested that everyone should reduce salt intake, others point out that there is no evidence that salt restriction is of any benefit to otherwise healthy individuals.[cxcix] *[This end note is listed in one of the three pages of "end-notes" at the end of the book but this point is so important that I also showed the reference site in the paragraph above.]*

SALT AND BLOOD PRESSURE
The overwhelming public interest in salt consumption derives from the concern over its perceived universal impact on blood pressure (BP). Unfortunately, this has long been a subject of significant myth-information. The cross-population blood pressure response to salt reduction is heterogeneous. With major

reductions in salt, about 30 percent of the population will experience a slight drop (2-6 mm) in systolic BP, while about 20 percent will see a similar increase in BP, and the remaining 50 percent of the population will show no effect at all. Considering the relatively small impact of major salt reduction on blood pressure, it is unfortunate that consumers are not aware of all the other negative consequences that occur as a result of dietary salt reduction.[cc]

Experiments are done with wrong salt.

First of all—scientific tests and experiments are performed with *refined* NaCl! That's right. Sea Salt or Natural Salt has never been used in the countless tests performed in the labs. This is due to fact that there are eighty some elements involved with *natural* sea salt. It would be statistically impossible to standardize a natural salt test due to its variance from one sample to the next.

In Vitro

The clarification we need here is that any experiments for salt affecting the body are always done **in vitro** (in test tube). No salt recommendation has ever come from facts derived from tests IN the human body. Almost all have been in test tubes with non-natural salt. A few tests have been performed on rats. The assumption is drawn that "If this thing happens here then this could happen over there". Maybe it does and then it might be different, since scientists come out with thousands of new reports from new findings every day.

In Vivo

In vivo refers to *"in" the body*. An "in vivo" test or experiment is performed on human tissue and to a living person. In vivo (Latin for "within the living") is experimentation using a whole, living organism.[cci]

Conclusion:

So, really, how much knowledge did we get from these experiments? We have received "some" information, yet nothing conclusive. We need answers without qualifying statements such as "might be a correlation". Is this information good enough and viable to make discernment for life altering and dangerous decisions? Our medical profession, government agencies and politicians may not really know the whole truth. They do have strong opinions.

Doctors are in charge of the "tests". They tell us what to do next by reading the results. However, we still have to live with the results. Be sensitive to our health, gain all the knowledge available and exercise discretion. Use moderation. Question everything.

Politicians deciding how much and what to eat assumes we are idiots. Unfortunately, many people are so busy that we do not educate ourselves to the extent we should when it comes to health. Why did the manufacturer produce the heavily processed food that has bad health benefits? Because we BUY the bad stuff! By making it the way they do, it is cheap, lasts forever and is profitable. We can only care for ourselves. When we only buy healthy product, they will only sell the same.

Chapter Twelve—History

As long as the foundation of earth itself, salt has existed on land and sea. It has played its role in all life. Since NaCl is in all soil, it is present in all living plant life. [ccii]

Salt is present in smaller amounts in fresh water and is 3.5% of sea water. All fish, reptiles, amphibians, birds, and mammals carry within their veins the element sodium.

So, to be clear, we are speaking of "mans" history with salt. However, it is so essential to life a few comments about salt is a must. Kingdoms and fortunes have risen and fallen around the ownership of scarce salt. As techniques were developed to acquire salt, it became more available and affordable. Today it is cheap but the "salt lore" is still around.

Perhaps because of the legend about spilled salt, Leonardo da Vinci's famous "Last Supper" has a spilled saltcellar in front of Judas. In the Kunsthistorische Museum in Vienna resides a magnificent sixteenth-century Golden Salt Cellar, product of the craftsmanship of Benvenuto Cellini.

Salt has played a prominent role in determining the power and location of the world's great cities. Liverpool rose from just a small English port to become the prime exporting port for the salt dug in the great Cheshire salt mines and thus became the entrepôt (trading post) for much of the world's salt in the 19th century.

Salt created and destroyed empires. The salt mines of Poland led to a vast kingdom in the 16th century, only to be destroyed when Germans brought in sea salt (which most of the world considered superior to rock salt). Venice fought and won a war with Genoa over salt. However, Genoese Christopher Columbus and Giovanni Caboto would later destroy the Mediterranean trade by introducing the New World to the market.[1]

Cities, states and duchies along the salt roads exacted heavy duties and taxes for the salt passing through their territories. This practice even caused the formation of cities, such as the city of Munich in 1158, when the then Duke of Bavaria, Henry the Lion, decided that the bishops of Freising no longer needed their salt revenue.

Solnitsata, the earliest known town in Europe was built around a salt production facility. Located in present-day Bulgaria, archaeologists believe the town accumulated wealth by supplying salt throughout the Balkans.

Aside from being a contributing factor in the development of civilization, salt was also used in the military practice of salting the earth by various peoples, beginning with the Assyrians.

It is commonly believed that Roman soldiers were at certain times paid with salt. (They say the soldiers who did their job well were "worth their salt.") The word 'salary' derives from the Latin word *salārium*, possibly referring to money given to soldiers so they could buy salt. The Roman Republic and Empire controlled the price of salt, increasing it to raise money for wars, or lowering it to be sure that the poorest citizens could easily afford this important part of the diet.

It was also of high value to the Hebrews, Greeks, Chinese, Hittites and other peoples of antiquity.

In the early years of the Roman Republic, with the growth of the city of Rome, roads were built to make transportation of salt to the capital city easier. An example was the Via Salaria (originally a Sabine trail), leading from Rome to the Adriatic Sea. The Adriatic Sea, having a higher salinity due to its shallow depth, had more productive solar ponds compared with those of the Tyrrhenian Sea, much closer to Rome.

During the late Roman Empire and throughout the Middle Ages salt was a precious commodity carried along the salt roads into the heartland of the Germanic tribes. Caravans consisting of as many as forty thousand camels traversed four hundred miles of the Sahara bearing salt to inland markets in the Sahel, sometimes trading salt for slaves: Timbuktu was a huge salt and slave market.[cciii]

The *gabelle*—a hated French salt tax—was enacted in 1286 and maintained until 1790. Because of the gabelles, common salt was of such a high value that it caused mass population shifts and exodus, attracted invaders and caused wars.

In American history, salt has been a major factor in outcome of wars. In the Revolutionary War, the British used Loyalists to intercept Revolutionaries' salt shipments and interfere with their ability to preserve food. During the War of 1812, salt brine was used to pay soldiers in the field, as the government was too poor to pay them with money. Before Lewis and Clark set out for the Louisiana Territory, President Jefferson in his address to Congress mentioned a mountain of salt supposed to lay near the Missouri River, which would have been of immense value, as a reason for their expedition. By 1810, new discoveries of salt along the Kanawha and Sandy Rivers greatly reduced the value of salt.

During India's independence movement, Mohandas Gandhi organized the Salt Satyagraha protest to demonstrate against the British salt tax.[cciv]

In the next chapter we will carry this theme of the history of salt a little further with salt mines and museums. The next chapter lists mines that are 7,000 years old and a mine that was fought for in the Civil war. I have also listed end-notes to look up for further information.

Chapter Thirteen—Salt Mines

This "read" on salt is not a coffee table top picture edition about salt mines but we should go over the basics of the fascinating world of salt mines since we are thinking about the subject of salt.

"Back to the salt mine"--- to return to one's work. Jocular; the image of menial and sometimes hard labor working in salt mines. We have all said it. Many of you mean it.

Though most salt in the world is produced by evaporation of sea water, there are still millions of tons of salt being produced each year from the mines deep in the ground.

Over 200 million tons of salt were produced in the world in 2006. China is the largest producer, with 48 million tons, followed closely by the United States, with 46 million tons.[ccv]
There are just 15 active salt mines in the United States.

Remember one important thing. All the salt produced has its origins in the sea, being either harvested directly from sea water or derived from rock salt deposits (former seas which have evaporated many millions of years ago).

Many rock salt deposits were formed over 200 to 250 million years ago as a result of the evaporation of earlier seas. A typical example is the deposits formed in the Zechstein Sea – these deposits were laid down 250 million years ago in a basin that stretched from the UK to Poland. More recently, around 16 million years ago, the Mediterranean Sea almost dried up, leaving large salt deposits that are now mined in Southern Spain.

Mines

Deep underground, cradled in the bosom of the earth, salt mines have a fascinating history and a remarkable current mystique. Use your imagination and open your mind while you experience the captivating underworld of salt mines.

Detroit Michigan area has one of the world's largest rock salt mines in the world. It is one of the world's largest salt mines. Miners at this mine worked at the unfathomable depth of 1,200 feet deep beneath Detroit. The Detroit salt mine, which consists of 100 miles of tunnels covers over 1,500 acres. It has never experienced a collapse or mine fatality. For years it produced tons of rock salt daily, most of which is used for ice control. Though the mining operations have ended, the mine still lurks right below the city.

Kansas Underground Salt Museum in Hutchinson, Kansas is unique in the western hemisphere. Although there are some similar salt mine museums in Europe, there are none in the Americas. The Kansas Underground Salt Museum's hours: 9AM - 5PM on Tuesday - Friday, 9AM - 6PM on Saturday, and 1-5PM on Sunday. Call 866 755-3450.

Louisiana has deep salt mines at Winn Parish and at Lake Peigneur, however, the Lake Peigneur site has experienced problems. A massive nine acre sink hole has opened up the landscape at this time and it is still growing. Another one seems eminent nearby...oops.

Cleveland Salt Mine is a world that most people will not see, however it has more than 400 miles of underground roads with rooms after rooms. Amazingly, the mine is all under the Great Lakes! That's right; the mine is under Lake Erie. To get to their jobs, workers at the Cleveland Salt Mine make a daily descent nearly 2,000 feet straight down. Most of the salt that's dumped on northern highways comes from ancient salt beds under the Great Lakes. The Cleveland salt mine produces almost 2 1/2 million tons of salt a year.

Saltville Virginia, has historical significance during the Civil War due to its salt flats which served as the site of one of the Confederacy's main salt works. The deposits left from millions of years past fed the wooly mammoths, the first humans, the Indians and the European settlers. It role in the American industry and the Civil War was pivotal. During the Civil War, salt was essential to preserving the provisions of the Confederate army and Saltville, mines supplied much of the salt for the country in 1863.

The Union Blockade made supplies scarce and the Confederacy was more and more dependent on Saltville. The Union army made several attempts to take the town and even with 5,000 men in 1864, led by Burbridge, he failed to come within sight of Saltville being protected by 2,800 Confederate soldiers and the Union retreated.

Wieliczka near Krakow, in southern Poland is listed as a world class attraction by the UNESCO World Heritage. The mine, built in the 13th century, produced table salt continuously until 2007, as one of the world's oldest salt mines still in operation. Tourists are amazed at the grandeur of the finely carved statues and wall carvings such as the Saint Kinga's Chapel in the Wieliczka Salt

Mine. The historic mine extends for a total of about 186 miles and functioned continuously since the Middle Ages until 1996.
The mine's attractions include dozens of statues, three chapels and an entire cathedral that has been carved out of the rock salt by the miners. The oldest sculptures are augmented by the new carvings by contemporary artists. About 1.2 million people visit the Wieliczka Salt Mine annually.

Your guided tour takes you from 210 to 440 feet underground, passing through galleries and chambers on three levels. The guided walk consists of 20 monumental chambers joined by 1.5 miles of pathways.

The Wieliczka salt mine is so impressive it has a history of inspiration. Some of the more famous inspirations are; he scenes in Bolesław Prus' 1895 historical novel, "Pharaoh" and in 1995, the Preisner's Music, by the Polish composer Zbigniew Preisner, *was recorded in the chapel* at the Wieliczka salt mine. The chapel is often referred to as having the best acoustics in Europe.

Hallstatt, Austria has the oldest salt mine in the world! It has been producing salt for 7,000 years. Situated in the mountain high above the village of Hallstatt in the Salzkammergut region of Upper Austria, Hallstatt provides a beautiful landscape. It is believed that Neolithic man
obtained salt from this mine; archaeological artifacts dating back to 5,000 BC has been discovered in the immediate vicinity.

The Hallstatt Salt Mine lays on the mountain a couple of hundred meters in height above the present-day village of Hallstatt. It is accessible by funicular railway for the first part of the ascent and then by hiking the last part of the journey.

To visit the museum you must first ride up to the mountain on the panorama funicular railway to trace the mysterious power of attraction of this mystical area. The path leads to the tunnel mouth and into a unique underground world: many miles of tunnels, which were started by hand by people more than 3,000 years ago, lead deep into the rock interior. The beams of light from the torches sweep over the naked, glistening rock face and briefly illuminate the places where hard-working miners once labored so very long ago.

The whole setting in Austria and near Germany and with other tombs and history, this site has high rave reviews for tourists.

Pakistan, Khewra salt mine, the second largest in the world, is a warren of 40 kilometers of tunnels housing an illuminated mosque made from salt rock, an electric train and even an asthma clinic.

The mine 100 miles south of Pakistan's capital Islamabad is the largest and oldest salt mine in the country, drawing up to 250,000 visitors a year.
Khewra was discovered back in 320 BC by Alexander the Great's troops, however. Trading did not start until the Mughal era in the sixteenth century.

Inside, the salt mine is so large, it has a mosque and electric railway where tourists can follow in the footsteps of Alexander the Great's troops. [ccvi]

Nemoco, Columbia. The five-century-old Nemocon mine is a fascinating look at the working conditions and daily lives of salt miners, such as their daily chapel attendance deep within the mine. More sights include a 3,500 lb. salt crystal carved in the shape of a heart, a wishing pool and a waterfall of salt. The Salt Museum, located in the town's oldest building just outside the mine, showcases the history of salt processing. A large chamber of mirrors and salt waterfalls highlighted with lights make a spectacular view.

Summary of Mines

Salt mines are clean, dry, and about 64*, so they make excellent storage facilities. Even the US Government uses salt mines extensively for storage. They also can be at times beautiful and definitely intriguing. I will leave with a *partial* list of some mine sites. There are many additional ones, of course. If you are doing a fast read, the following listing should furnish enough suggestions about salt mine locations throughout the world.

Rheinburg, Germany

Slanic, Cacica, Mari, Salina Turda, Ocna, and Praid....in Romania

Provadiya, Bulgaria

Racalmuto, Realmonte, Petralia Soprana....Italy

Cheshire, Worcestershire....England

Goderich, Canada

Khemisset, Morocco

Tuzla, Bosnia

Africa and West Africa

Chapter Fourteen—Foods with Salt

Seaweed, green leafy vegetables and tuberous vegetables are natural sources of sodium (salt), as well as other essential salts, such as magnesium and potassium

Foods to Be <u>Avoided or limited</u>: [High in refined, over-processed Salt]

- Processed foods
- Convenient foods
- Fried or smoked
- Hamburgers
- Pizza
- Chicken nuggets
- Meat pie
- Sausage
- Pastries
- Cakes
- Noodles, canned chicken noodles
- Breakfast cereals
- Bacon
- Tortilla
- Chips
- Popcorn (if salted)
- French fries
- Pretzels...or any salted cracker
- Cheese (one small slice has 160 mg)
- Beverages (processed, canned, bottled)

- Bread
- Condiments like ketchup, salad dressings, sauces, mustard, dips,
- All packaged or canned food including soups
- Lunch meats and cured meats
- **Don't** even think about salt substitutes!

Natural, Healthy foods with Salty Taste

- Lemon juice
- Garlic powder
- Onion powder
- Lime juice

Of course, there is the natural, un-processed salt your body craves in order to be healthy......Sea Salt!

Chapter Fifteen— Ocean

Since salt plays a primary function of the sea, we need to expound on the roles of this abundant presence of salt. The oceans and its salt are the life line of the planet. So, "salt" in the ocean deserves some attention.

~~~~~Imagine you are cruising somewhere on a blue tropical ocean. Looking all around every view you see is a beautiful watery paradise. Earth is a watery planet, after all. Your body biologically feels the healthy ions flowing up from the sea, energizing you and repairing the electrical impulse system damaged by the pollutants we are constantly being bombarded with. For a moment, lets reflect on how the ocean, with its strong but sensitive system works to supply and heal the planet.

We already know that the oceans are 98% of the earth's water and covers 70% of the Earth's surface. The water is made up of 3.5% salt that flowed from the freshwater rivers and streams that were only .5% salt. The fresh water becomes slightly acidic by combining with the carbon in the air and this carbonic acidity is able to dissolve minerals. The water then gathers its salt as the rain water percolates through the soil and rocks dissolving some of the minerals. This *weathering* provides some but not all of the salt in the sea. The sea has pretty much stabilized for salinity but acquires its supply of salt like this:

- Rivers and streams weather the rocks and deposit salt and minerals
- Hydrothermal vents on the ocean floor heats up the floor crust and by dissolving the minerals and salts it deposits them back into the ocean
- Submarine volcanoes erupt all the time on the ocean floor and deposits salts into the water
- The two most abundant ions in seawater are chloride and sodium and make 90% of all dissolved ions in the sea.
- All of the salt in the oceans would cover the Earth's land 500 feet thick.

Salts role in sea currents

The salt in the water makes the salty water heavier than less salty water, helping to create the thermohaline circulation system known as the Global Conveyor Belt. The heavy salty water sinks down creating circulation acting as a submarine river that moves water throughout the ocean. The seventeen major currents found in the world's oceans affect the weather, animal life in the ocean and therefore the Earth's balance. Some of the major currents are; California Current, Humboldt Current, Gulf Stream, and the Labrador Current.[ccvii]

Density differences are a function of temperature and salinity and provide the pumping action to drive 90% of the ocean through "trade" currents. The Gulf Stream for example is a warm current that originates in the Gulf of Mexico and moves north toward Europe. The Humboldt Current is another example of a current that affects weather. When this cold current is normally present off the coast of Chile and Peru, it creates extremely productive waters and keeps the coast cool and northern Chile arid. However, when it becomes disrupted, Chile's climate is altered and it is believed that El Nino plays a role in its disturbance. The Labrador Current, which flows south out of the Arctic Ocean along the coasts of Newfoundland and Nova Scotia, is famous for moving icebergs into shipping lanes in the North Atlantic.[ccviii]

Salt keeps the oceans alkaline

Salt helps keep the acidity down in the ocean. This is especially important now because the acidity of the ocean appears to be up slightly. Scientists think that the acidity is up due to dead plankton, water contamination and salt being diluted with melted ice. This acidity prohibits the salty sea water from absorbing more CO2. So, we need salt in the sea.

Salt water also keeps the ocean alkaline at 8 ph. This is important to keep the plankton alive. Plankton not only provides enormous amounts of food for other wild life like whales, it is the major source of oxygen for the planet! No plankton—no life for any of us.

Salt water absorbs large amounts of CO2 and the oceans are by far the largest carbon sink in the world. Some 93 percent of carbon dioxide on the earth is stored in algae, vegetation, and coral under the sea making it the biggest way to balancing the air. The ocean holds 50 times more carbon than the atmosphere.

Coastal areas absorb large amounts of carbon!

Coastal marine ecosystems could become instrumental in mitigating climate change concerns according to the International Union for the Conservation of Nature (IUCN) and the ESA (Ecological Society of America). The carbon sequestered in vegetated ecosystems, specifically—mangrove, sea grass beds and salt marshes has been termed "blue carbon". Though their global area is one to two orders of magnitude smaller than that of terrestrial forests, the contribution of vegetate coastal habitats per unit area to long-term C sequestration is much greater, in part because of their efficiency in trapping suspended matter. [ccix]

"If you look at the quality of carbon compared to forests you will find these habitats are 15 times more effective per unit area," said Dan Laffoley, vice chair of IUCN's World Commission on Protected Areas and an author of the report. "This has been a big wake-up call."

The United Nations estimated in a report that 3-7 percent of current fossil-fuel emissions could be offset in two decades if more action is taken to prevent marine vegetation loss and degradation worldwide. It currently absorbs more than 50% of all fossil-fuel emissions.[ccx] Sounds more realistic than the popular expensive remedies we are working on now.

Reminder; Carbon only contributes 2% of any warming affect in the atmosphere. That is all the carbon in the air, no matter where it is originated from. This mentioning of carbon is in reference to *concerns* of carbon.

Satellite to observe the salinity of the ocean

Launched June 10, 2011, aboard the Argentine spacecraft Aquarius/Satélite de Aplicaciones Científicas (SAC)-D, Aquarius is NASA's first satellite instrument specifically built to study the salt content of ocean surface waters. Salinity variations, one of the main drivers of ocean circulation, are closely connected with the cycling of freshwater around the planet and provide scientists with valuable information on how the changing global climate is altering global rainfall patterns.

Aquarius will provide the first global observations of salinity covering Earth's oceans once every 7 days, collecting as many sea surface salinity measurements as the entire 125-year historical record from ships and buoys with an accuracy of about a "pinch" of salt in 1 gallon of water.

Since 86% of water evaporation and 78% of all rain fall occur over the oceans, the salinity in the ocean varies. This changing salinity affects the ocean dynamics and can now finally be studied with the use of the satellite, Aquarius.[ccxi]

Chapter Sixteen— Variety of Salts

Table salt

Table salt is the most common salt found in most kitchens. It's a finely ground, refined form of rock salt, slightly bitter tasting from additives used to keep it from clumping. [ccxii]Most minerals are removed during processing. Some forms of table salt are artificially treated with iodine.

Iodized table salt contains a small amount of potassium iodide and dextrose (a sugar used to stabilize the iodide) as a dietary supplement to prevent goiter and mental retardation.

All table salt contains an anti-caking agent like calcium silicate, aluminosilicate (aluminum and silicon) and yellow prussiate of soda. These keep it from clumping in humid conditions so it flows freely from the box.

Gourmet Salts

Many natural salts have gained gourmet status; and, are recommended for their rich mineral content providing numerous health benefits. Choosing a salt depends on individual taste preferences and upon the application for which it will be used. So without further ado -- I give you salt.

Gourmet sea salts vary based on how they are harvested and the extent to which they are refined. Some coarse or fine sea salts are about the same in composition as regular table salt, containing 99% sodium chloride and 1-2% magnesium and calcium chlorides and other trace minerals. Varieties like *sel gris* ("gray salt") are a moist salt that is not refined, so it contains clay and other trace elements from the evaporation ponds. The most premium of all sea salts, fleur de sel ("the flower of salt"), consists of delicate crystals skimmed from the surface of the evaporation pond by hand. Gourmet sea salts sell for $20-$50 per pound or more.

Exotic sea salts like Hawaiian red salt and Indian black salt contain clay, which gives them their unique color and flavor.

Kosher Salt

Kosher salt has a milder, less pungent taste than table salt and is the choice of many chefs. The flavor disperses quickly as it dissolves fast. The course crystals are excellent for curing meats.

This salt is a coarse, flaky salt. It is not iodized, and depending on the brand it may or may not contain an anti-caking agent like Yellow Prussiate of Soda (sodium ferrocyanide).

Kosher salt is produced using two methods. The industry standard method used by Morton Salt is to flatten salt crystals into flakes using rollers. Cargill, maker of Diamond Crystal Kosher Salt, uses a method called the Alberger process. A brine solution is heated in an 80' x 40' open vat. Large rakes agitate the steaming brine, and as it evaporates, crystals form into tiny pyramids with jagged edges. Cargill claims their kosher salt dissolves faster and clings to food better than rolled kosher salt.

Celtic Salt

Celtic sea salt comes from the northwest coast of France, where it is harvested after seawater evaporates from clay ponds built near the shoreline. Live minerals and trace elements provide natural gray Celtic sea salt with its salty flavor and health benefits such as enhanced digestion, resistance to infections, kills bacteria and increases alkaline levels.

According to Juicing for Health, the magnesium content of Celtic sea salt "ensures that unused sodium is quickly and completely eliminated" from the body through the kidneys. Its abundant minerals extract excess acidity and balance electrolytes vital to cells processing communication and information in the brain.

Celtic sea salt minimizes water retention while stimulating saliva and increasing the drainage and elimination of mucus. "It is important that you increase your water intake when taking Celtic salt," says Juicing for Health. Your body is very quickly eliminating toxins, and water is needed to flush them from your system. Celtic sea salt also helps build the immune system, boost energy and balance blood sugar in diabetics. Its minerals guard against muscle cramps.

Celtic sea salt assists the healing process during illness, burn and post-surgery situations by providing potent minerals directly to the body's cells. It also helps heal adrenal and thyroid disorders, headaches, skin diseases, high cholesterol levels, cataracts and congestion of the sinuses. A natural antihistamine, it relieves allergies. Celtic sea salt also may have a role in the treatment of cancer.

According to CureZone; "In the theory of acid and alkaline balance, chronic disease such as cancer is caused by the acidification of the blood, lymph and all cellular tissues. Real sea salt is one of the basic elements necessary to correct this problem." [ccxiii] Unrefined salt with its minerals naturally alkalizes the body.

Course Salts

Himalayan Pink Salt: A star among salts, Himalayan salt is typical of coarse salts, with large-grained crystals best used in a salt grinder. Course salts are not as moisture sensitive as other types, allowing them to be stored for long periods. Himalayan pink salt is unrefined and high in minerals, making it a healthful choice. It is useful for both seasoning or as a finishing salt.

Seasoning Salts

Kala Namak is an unrefined, authentic Indian salt with a strong sulfuric flavor. It is preferred by vegan chefs for adding an egg-like flavor to dishes.

Hawaiian Alaea sea salt

Traditional red-colored salt used for preserving and seasoning foods. This salt is enriched with *Alae*, volcanic baked red clay, which adds iron oxide for color and flavor. Earthy and mellow tasting and used in authentic Hawaiian dishes.

French sea salt

Hand harvested from the Atlantic coast of France, this salt is unrefined and high in minerals, especially natural iodine. French sea salt is a perfect replacement for the chemical taste of iodized salt. The salt has a moist texture and is lower in sodium chloride than other salts.

Finishing Salts

Salt draws out and enhances the flavor of food during cooking. *Finishing salt* is sprinkled just before eating, "...adding a burst of salty goodness and crunchy texture at the very end".

- Italian sea salt

From the coast of Sicily, this unrefined salt is rich in magnesium, iodine, fluorine, potassium and sodium chloride. It is considered delicate and flavorful.

- Hawaiian Hiwa Kai or Black sea salt

This salt is black in color due to the addition of activated charcoal, which enhances the flavor. The charcoal is known for its ability to aid in detoxing, neutralizing stomach acids and helping to prevent acid reflux.

- Celtic sea salt or Grey salt

Grey salt, collected by hand on the Brittany coast in France is considered one of the best salts by many in the culinary world. The unrefined salt is loaded with minerals, comes in coarse, fine or extra fine grind and provides a rich, luscious flavor.

- Fleur de sel

Considered the caviar of salts, this specialty salt is hand harvested from the Guerande region salt ponds in France. The salt blooms like a flower on the water's surface under "just right" weather conditions. It's only harvested once a year. It is said to melt slowly on the tongue with a lingering, earthy flavor.

- Smoked salt

Smoked salt is an aromatic edible salt product with smoke flavoring. It is a seasoning and is used as a shortcut to add smoked flavor to foods. Smoked salt consists mainly of sea salt and smoke volatiles condensed on the salt. An ingredient typically listed on smoked salt is sawdust.

Smoked salt lends a strong smoke flavor and aroma to meats and vegetables, and is suitable for vegetarians. "Smoked salt" differs from "smoke flavored salt" in that the latter contains a smoke flavored additive and is therefore not classified as a pure salt product. Smoked salt is typically made from evaporated "sea salt" as opposed to "mined salt". Both types of salt are sodium chloride (NaCl), but mined salt often has a more pronounced "salty" taste due to the iodine added for health reasons; therefore mined salt is typically not suitable for use in the manufacture of smoked salt. Smoked salt is a manufactured product. There are several ways to manufacture it:

1. Place salt in a wood smoker.
2. Coat salt with smoke-flavored oil(typically referred to as liquid smoke)
3. Mix salt with smoke-flavored maltodextrin[ccxiv]

Sprinkling Salts

Most restaurant and professional chefs are taught to salt a dish by using the *three-fingered pinch* method. Although this seems like a lot of salt the first time you take a pinch, in reality it's the equivalent of 1/8 to 1/4 tsp. So salt with style using this method and stay in control to prevent over-salting a dish. What's in your salt cellar?

*Bath Salts—*Not a salt*

Bath salts is the informal "street name" for a family of designer drugs often containing substituted cathinones, which have similar effects as amphetamine and cocaine. Their white crystals resemble legal bathing products like Epsom salts, but are chemically disparate from actual bath salts. Bath salts' packaging often states "not for human consumption" in an attempt to avoid the prohibition of drugs. Other "street names" for this drug are Ivory Wave, Purple, Vanilla Sky and Bliss.

Salt Substitutes—(Low salt and "no salt" food)

Salt Substitutes are 100% potassium chloride. They contain no sodium chloride. Salt substitutes are used by people on low-salt diets, in processed food labeled "low-salt" and mixed up consumers. _Potassium chloride_ is a by-product when making nitric acid from hydrochloric acid and does not belong in your body. It greatly affects your electrolytes and too much will stop your heart cold. Dead.

Potassium chloride, can taste like salt and is an electrolyte, yet does not affect your body in the same manner as sodium chloride. However, potassium is not always a safe substitute. Excess potassium in your diet can affect you if you have kidney problems.

The fear is, as food processors try to create "salt free" and "low salt", they simply use the toxic substance, "_potassium_" _chloride_ in its place. [BTW: same goes for _low sugar_ food. It is magic is it not? They sell us food that tastes good and it has no sugar! No. Sorry. They use sugar substitutes that are many times worse for your health and actually raise your sugar index more than plain sugar]

Chapter Seventeen—Salt Uses

Salt for air ionizer

Salt can function as an air ionizer, actually releasing negative ions into the surrounding air. The air we breathe is filled with millions of particles, molecules, ions and living organisms. These particles and toxins are removed with air ionizers.

One type of air ionizers are salt crystal lamps. In the presence of heat, negatively charged ions disperse into the air and attract the dust particles and pollutants which are positively charged. The heavier substance then sinks to the floor, thus cleaning the air. A good example of a salt ionizer is the Himalayan salt crystal lamp. [ccxv]

Anti-bacterial

~Killing bacteria

Salty water causes the cell of bacteria to dehydrate, which eventually kills the cell. Bacteria is a common skin inhabitant. Your skin tends to be salty, so this is one way your body protects you against bacteria on your skin.

Some bacteria, such as Staph, have adapted to living in salty environments, but even Staph can't live in highly salty surroundings, such as salted foods like ham.

The major effect of salt as a preservative is that it withdraws water from microorganisms if the external salt concentration is high enough. The microbes would shrivel and die. [ccxvi]

~Salt in your wound
Sterile saline solution (0.9% sodium chloride) can be used to effectively clean any type of wound, no matter how deep. You can find saline solution for wounds in the first aid section of a pharmacy. Because this solution matches the pH of your body fluids and blood, it does not burn at all when applied to wounds. ccxvii

~IV (intravenous therapy)
In hospitals, saline solution (again, the 0.9% sodium chloride) is used as a hydrating IV (intravenous therapy) fluid as well as irrigation for bone-deep sores, so you can rest assured that it is gentle on your body!

Seasoning

*Never use table salt
*Use kosher salt for general seasoning and cooking
*Use sea salt, especially Fleur de Sel, strictly for finishing a dish
*Experiment with different colored and flavored salts to experience new tastes and textures.
*Cook with Passion

Water softening

The calcium and magnesium in water both carry positive charges. This means that these minerals will cling to the beads in the special mineral tank in your water softener as the hard water passes through.

Sodium (salt-NaCl) ions also have positive charges. Although sodium does not have as strong a charge as the charge on the calcium and magnesium, when a very strong brine solution is flushed through a tank that has beads already saturated with calcium and magnesium, the sheer volume of the sodium ions is enough to drive the calcium and magnesium ions off the beads. Water softeners have a separate brine tank that uses common salt to create this brine solution.

In normal operation, hard water moves into the mineral tank and the calcium and magnesium ions move to the beads, replacing sodium ions. The sodium ions go into the water. Once the beads are saturated with calcium and magnesium, the unit enters a 3-phase regenerating cycle. First, the backwash phase reverses water flow to flush dirt out of the tank. In the recharge phase, the concentrated sodium-rich salt solution is carried from the brine tank through the mineral tank. The sodium collects on the beads, replacing the calcium and magnesium, which go down the drain. Once this phase is over, the mineral tank is flushed of excess brine and the brine tank is refilled.[ccxviii]

Curing Meat

Salt-cured meat or salted meat, like bacon and kippered herring, is meat or fish preserved or cured with salt. Salting, either with dry salt or brine, was the only widely available method of preserving meat until the 19th century. It was frequently called "junk" or "salt horse".

Salt inhibits the growth of microorganisms by drawing water out of microbial cells through osmosis. Concentrations of salt up to 20% are required to kill most species of unwanted bacteria.

Salted meat and fish are a staple of the diet in North Africa, Southern China, Scandinavia, coastal Russia, and in the Arctic. Salted meat was a staple of the mariner's diet in the Age of Sail. It was stored in barrels, and often had to last for months spent out of sight of land. The basic Royal Navy diet consisted of salted beef, salted pork, ship's biscuit, and oatmeal, supplemented with smaller quantities of peas, cheese and butter.

Salt beef in the UK and Commonwealth as a cured and boiled foodstuff is sometimes known as *corned beef* elsewhere, though traditional salt beef is different in taste and preparation. The use of the term *corned* comes from the fact that the Middle English word *corn* could refer to grains of salt as well as cereal grains.[ccxix]

Melt ice on roads

Even though salt may be applied dry it does not begin its snow-fighting job until it dissolves into brine. A chemist would explain the process in terms of colligative properties. The brine is a solute and the concentration of grains in the solute (in this case, salt brine) determines its freeze-point lowering potential. Any substance that dissolves in water has this effect, but each substance will have varying outcomes. Salt has a lower molecular weight and gives it almost six times the effectiveness of say sugar in lowering the freezing point of water – actually even more in this example since sugar isn't an electrolyte at all. This is the same principle you use when you put antifreeze into your car's radiator.

Salt applied as a liquid or pre-wet solid can begin to act immediately lowering the freezing point of water. On a pavement where the temperature is 30°F (-1° C), one pound of salt melts 46.3 pounds of ice. One inch of ice on one lane-mile of road would weigh 70 tons. To melt that much ice would take 17 tons of salt. But the objective is not to melt the snow and ice off the pavement, only to prevent or destroy the bond on the surface of the roadway between the pavement and the ice or snow. In our example lane-mile with an inch of ice, most road agencies would use 500 pounds or less, less than 2% of the amount of salt needed to melt the ice.

The objective being to prevent the bond if possible (not melt all the ice), liquids are appropriate when applied in a pre-storm anti-icing application to be in place before freezing precipitation arrives. It also explains why agencies use larger particles for application of dry salt to ice- and snowpack-covered roads since they need to have the weight and mass to bore down to the pavement where the real work is done.[ccxx]

*"Salt Peter" is not a salt

Salt peter typically refers to the chemical compound potassium nitrate, though it may also refer to sodium nitrate. Salt peter was once collected from decomposing material, but today, it is generally manufactured by treating sodium nitrate, mostly mined in Chile, with potassium chloride and collecting the precipitate. Salt peter was one of the ingredients of the first gunpowder, black powder. Today, it has many uses in both the laboratory and the larger world.

Black powder, oxidized with salt peter, is still used for small novelty explosives, such as fireworks and model rockets, though firearms typically use newer types of gunpowder. Saltpeter is most widely used in manufacturing nitric acid, however, which is in turn used to make Trinitrotoluene (TNT) and other modern explosives.

Explosives

- An easy **explosion** can be performed from **salt**, aluminum, and flash source.
- Make Gunpowder with Salt and Sugar[ccxxi]
- With only some sugar, **salt** substitute and an instant cold pack, you can **make** your very **own explosive** gas!
- Burn saltwater for fuel...National Geographic (possible? Take a grain of salt and read the claim below)

According to an article in the *Pittsburgh Post-Gazette,* salt water can indeed burn when exposed to a certain kind of *radio wave*, a university chemist has confirmed. Rustum Roy of Pennsylvania State University verified earlier this month that the radio waves break the water into its components, allowing the resulting freed hydrogen and oxygen to catch fire.

When he trained the radio waves on a test tube of salt water, it produced an unexpected spark, according to the *Pittsburgh Post-Gazette*.

Curious, they decided to ignite the water with a match. The water lit and kept burning as long as it remained in the radio frequency field. Roy (Pennsylvania State University) then further experimented for the technology's potential applications for desalination and hydrogen fuel.

He found that the phenomenon works by breaking water into oxygen, hydrogen, and salt. The hydrogen is combustible and will burn as long as it remains within the radio frequency field.
As salt water passes through the generator, the hydrogen would be released.
For now, the most immediate potential technology application is desalination—the process of removing salt from water—because the water formed after combustion is free of salt and other contaminants.
"It's really a miraculous process: water-breakup-water," Roy said. [ccxxii]

*[My Opinion...] Might make a good movie plot. Keep trying Roy.

Chapter Eighteen—SALT OF THE EARTH

We can't leave without citing an iconic phrase used throughout time.

Webster Dictionary: "Salt of The Earth"—person or persons of a group regarded as the **finest**.

Bible: Mathew 5:13 *Ye are the salt of the earth: but if the salt have lost his flavor, wherewith shall it be salted? It is thenceforth good for nothing, but to be cast out, and to be trodden under foot of men.*

Wikipedia: The role of salt in the Bible is relevant to understanding Hebrew society during the Old Testament and New Testament periods. Salt is a necessity of life and was a mineral that was used since ancient times in many cultures as a seasoning, a preservative, a disinfectant, a component of ceremonial offerings, and as a unit of exchange. The Bible contains numerous references to salt. In various contexts, it is used metaphorically to signify permanence, loyalty, durability, fidelity, usefulness, value and purification.

Salt has a special meaning that we all know. Defining it is a little more difficult to pin down. The many accolades like praiseworthy, reputable, worthy of respect, deserving, exemplary, honorable, laudable, suitable, believable, constant, favored, upright, illustrious, honored, straight, reliable, truthful, excellent, credible, righteous, fair, distinguished, dependable, best, decent, unpretentious, good, priceless, pure, virtuous, valuable and admirable, show how basic and important we realize it's role is.

Salt changes the taste and flavor of the substance it is mixed with. That fact is usually the only thought we give salt. We take for granted this tiny little ordinary substance. But the body can't live without it and the earth relies on it. It is vitally important to our existence.

Yet, in *my* mind, salt is much like the idea of "light" and in a sense… "knowledge". These wonderful traits are what make individuals truly interesting. Persons pervaded with the insatiable thirst for learning, add the flavor and the essence to civilized society. This is who *you* are and thanks for being special.

It was fun talking to you. Whether you agree on everything here, you are reading, exploring and thinking.

Pass the salt.

---

responsibility for the use or misuse of this material. Your use of this written material indicates your agreement to these terms and those published here. All trademarks, registered trademarks and service-marks mentioned in this book are the property of their respective owners.

---

[i] *http://www.education-world.com/a_issues/issues148a.shtml*

[ii] *http://www.webmd.com/rheumatoid-arthritis/news/20100528/rheumatoid-arthritis-is-on-the-rise*

[iii] *http://thyroid.about.com/od/symptomsrisks/a/People-Getting-Thyroid-Disease.htm*

[iv] *http://www.vs.gov.bc.ca/stats/quarter/q2_01/xl/chart1.xls*

[v] *According to Scientific American, June 2000, pg 30 "Asthma Worldwide",*

[vi] *http://news.bbc.co.uk/1/hi/programmes/newsnight/2224126.stm*
http://news.bbc.co.uk/1/hi/programmes/newsnight/2232111.stm
http://www.sundayherald.com/21347
http://www.vaccinationnews.com/DailyNews/March2002/ExpertSaysMMRWillBeProved.htm

[vii] *http://www.hindu.com/thehindu/holnus/003200609130358.htm*

[viii] *http://www.cdc.gov/ncipc/factsheets/suifacts.htm*

[ix] *According to WebMD, "There has been a 10-fold increase in the number of children with Type 2 diabetes during the past five years."*

[x] *Dr. Batmanghelidj's book, "Water: Rx for a Healthier Pain-Free Life. published in Hypertension: Journal of the American Heart Association."*

[xi] *Hhttp://www.pbs.org/newshour/rundown/2011/08/obesity-ratings-rising-worldwide-us-could-hit-50-by-2030html*

[xii] *discoverysedge.mayo.edu/celiac-disease/*

[xiii] *http://thechart.blogs.cnn.com/2010/09/02/americans-rx-drug-use-on-the-rise/*

[xiv] *http://www.health.harvard.edu/healthbeat/HEALTHbeat_062106.htm*

[xv] *http://christinecronau.com/if-margarine-is-healthier-why-wont-ants-eat-it/*

[xvi] *http://wellnessmama.com/2193/why-you-should-never-eat-vegetable-oil-or-margarine/*

[xvii] *http://www.huffingtonpost.com/2013/03/30/health-benefits-of-eggs-yolks_n_2966554.html*

[xviii] *http://www.blisstree.com/2012/07/19/food/study-sports-drinks-are-bad-for-you-968/*

[xix] *http://www.livestrong.com/article/457257-what-are-the-dangers-of-saccharin-aspartame-as-fake-sugars/*

[xx] *http://www.thirdage.com/nutrition/dangerous-side-effects-artificial-sweeteners*

[xxi] *http://www.alkalizeforhealth.net/Ldnadamage.htm*

[xxii] *http://christinecronau.com/if-margarine-is-healthier-why-wont-ants-eat-it/*

[xxiii] *http://articles.mercola.com/sites/articles/archive/2011/11/06/aspartame-most-dangerous-substance-added-to-food.aspx*

xxiv *http://organichealthadviser.com/archives/aspartame-dangers*

xxv *http://www.lowcarb.ca/articles/article146.html*

xxvi *http://www.thirdage.com/nutrition/dangerous-side-effects-artificial-sweeteners*

xxvii *http://www.webmd.com/healthy-aging/omega-3-fatty-acids-fact-sheet*

xxviii http://lpi.oregonstate.edu/infocenter/othernuts/omega3fa/

xxix *http://www.naturalnews.com/010095.html*

xxx *http://www.naturalnews.com/029194_cancer_risk_fats.html*

xxxi *http://www.doctoroz.com/blog/susan-evans-md/dangers-tanning-beds*

xxxii *http://www.examiner.com/article/how-much-does-congressional-pay-and-benefits-cost-taxpayers*

xxxiii *http://useconomy.about.com/od/usfederalbudget/p/US-Government-Federal-Budget-FY2012-Summary.htm*

xxxiv *http://www.oneresult.com/articles/nutrition/unhealthy-facts-about-fried-foods*

xxxv *http://www.healthy-food-site.com/burnt-food.html*

xxxvi *http://www.mnn.com/food/healthy-eating/questions/is-eating-burned-food-bad-for-you*

xxxvii http://www.livescience.com/31995-how-do-wind-turbines-kill-birds.html

xxxviii *http://logging.about.com/od/Eco-Friendly-Logging/tp/Carbon-Storage-In-Lumber-And-Trees.htm*

xxxix *http://www.livestrong.com/article/47899-pros-cons-recycling-plastic/*

xl *http://listverse.com/2013/01/27/10-ways-recycling-hurts-the-environment/*

xli *http://abcnews.go.com/WN/wind-power-equal-job-power/story?id=9759949#.UW7f6aKsh8E*

xlii *http://wizbangblog.com/content/2010/02/14/climate-scientist-phil-jones-no-global-warming-since-1995.php*

xliii *www.drroyspencer.com/2013/03/global-microwave-sea-surface-temperature-update-for-feb-2013-0-01-deg-c/*

xliv *www.drroyspencer.com/2013/03/global-microwave-sea-surface-temperature-update-for-feb-2013-0-01-deg-c/*

xlv *http://news.nationalgeographic.com/news/2009/07/090731-green-sahara_2.html*

xlvi *http://www.breitbart.com/Big-Peace/2012/12/20/Russians-Freeze-to-Death-in-Worse-Winter-Since-Stalin-s-Great-Purge*

xlvii *http://www.boston.com/bigpicture/2012/02/extreme_cold_weather_hits_euro.html*

xlviii *AccuWeather.com Long Range Expert Joe Bastardi believes there is a significant chance for particularly frigid winters in 2012-2013 and 2013-2014 into 2014-2015.*

xlix *http://iceagenow.com/New%20Little%20Ice%20Age.htm*

[li] *www.epa.gov/climatechange/science/indicates/ghg/global-ghg-emissions.html*

[li] *www.epa.gov/climatechange/science/indicates/ghg/global-ghg-emissions.html*

[lii] *http://rense.com/general75/0223_inconvenient_gore.pdf*

[liii] *RaRouchepac.com*

[liv] *http://www.theregister.co.uk/2013/04/22/climate_sensitivity_down_down/*

[lv] *http://www.aip.org/climate/oceans.htm*

[lvi] *http://en.wikipedia.org/wiki/Carbon_dioxide_in_Earth's_atmosphere*

[lvii] *http://www.humanesociety.org/assets/pdfs/farm/hsus-fact-sheet-greenhouse-gas-emissions-from-animal-agriculture.pdf*

[lviii] *http://www.humanesociety.org/assets/pdfs/farm/hsus-fact-sheet-greenhouse-gas-emissions-from-animal-agriculture.pdf*

[lix] *http://www.nsf.gov/news/news_summ.jsp?cntn_id=110580*

[lx] *http://wiki.answers.com/Q/What_percentage_of_CO2_emissions_come_from_human_activity*

[lxi] *http://www.sciencedaily.com/releases/2011/03/110316084907.htm*

[lxii] *http://rense.com/general75/0223_inconvenient_gore.pdf*

[lxiii] *William Davis author "Wheat Belly"*

[lxiv] *http://www.celiac.com/articles/22252/1/Celiac-Disease-Diagnoses-On-the-Rise/Page1.html*

[lxv] *William Davis author "Wheat Belly"*

[lxvi] *William Davis author "Wheat Belly"*

[lxvii] *http://www.dailymail.co.uk/femail/article-461095/The-sweet-truth-Ditch-sugar-look-years-younger.html*

[lxviii] *www.aging-no-more, inhuman experiment*

[lxix] *http://wellnessmama.com/2193/why-you-should-never-eat-vegetable-oil-or-margarine/*

[lxx] *www.aging-no-more, inhuman experiment*

[lxxi] *http://www.oliveoiltimes.com/olive-oil-health-benefits*

[lxxii] *http://www.marksdailyapple.com/defending-olive-oils-reputation/#axzz2UykdNzVk*

[lxxiii] *www.hazards of milk*

[lxxiv] *www.hazards of milk*

[lxxv] *http://www.scientificamerican.com/article.cfm?id=experts-organic-milk-lasts-longer*

[lxxvi] *http://www.naturalnews.com/030483_organic_milk_fraud.html#ixzz2U24Hf9o9*

[lxxvii] *http://www.naturalnews.com/030483_organic_milk_fraud.html*

[lxxviii] http://www.anh-usa.org/the-great-organic-deceivers/

[lxxix] http://www.naturalnews.com/030483_organic_milk_fraud.html

[lxxx] *http://www.smart-publications.com*

[lxxxi] http://www.knowthelies.com

[lxxxii] http://www.ahealedplanet.net/mdaq.htm

[lxxxiii] http://www.knowthelies.com

[lxxxiv] http://www.knowthelies.com

[lxxxv] www.wordpress.com

[lxxxvi] Grain.org

[lxxxvii] prof77.wordpress.com/.../monsantos-revolving-door-into-the-**fda-us**...

[lxxxviii] http://musicians4freedom.com

[lxxxix] http://www.jonbarron.org/article/pillar-salt

[xc] http://online.wsj.com/article/SB10001424127887324880504578298271512023796.htm l

[xci] http://articles.mercola.com/sites/articles/archive/2011/10/28/cdc-director-arrested-for-child-molestation--bestiality.aspx

[xcii] www.naturalnews.com/salt.html

[xciii] http://www.nytimes.com/2012/06/03/opinion/sunday/we-only-think-we-know-the-truth-about-salt.html?pagewanted=all&_r=0

[xciv] http://www.naturalnews.com/036946_salt_diet_myths_hypertension.html#ixzz2OgAHv Bhn

[xcv] http://www.westonaprice.org/vitamins-and-minerals/salt-and-our-health

[xcvi] http://summaries.cochrane.org/CD004022/effects-of-low-salt-diet-on-blood-pressure-hormones-and-lipids-in-people-with-normal-blood-pressure-and-in-people-with-elevated-blood-pressure#sthash.86TR54dn.dpuf

[xcvii] http://www.nytimes.com/2011/05/04/health/research/04salt.html

[xcviii] http://www.ncbi.nlm.nih.gov/pmc/articles/PMC126303/

[xcix] http://www.saltinstitute.org/Issues-in-focus/Food-salt-health/Salt-and-cardiovascular-health

[c] http://www.nhs.uk/news/2011/07July/Pages/heart-risk-salt-reduction-cochrane-review.aspx

[ci] http://www.nytimes.com/2012/06/03/opinion/sunday/we-only-think-we-know-the-truth-about-salt.html?pagewanted=all&_r=0

[cii] http://www.saltinstitute.org/Issues-in-focus/Food-salt-health/Salt-and-cardiovascular-health

[ciii] http://www.nytimes.com/2012/06/03/opinion/sunday/we-only-think-we-know-the-truth-about-salt.html?pagewanted=all&_r=0

[civ] http://www.ncbi.nlm.nih.gov/pmc/articles/PMC2118645/

cv http://www.nytimes.com/2012/06/03/opinion/sunday/we-only-think-we-know-the-truth-about-salt.html?pagewanted=all&_r=0

cvi U.S. News and World Report .

cvii http://www.heart.org/idc/groups/heart-public/@wcm/@hcm/documents/downloadable/ucm_300625.pdf

cviii http://www.westonaprice.org/vitamins-and-minerals/salt-and-our-health

cix http://www.westonaprice.org/vitamins-and-minerals/salt-and-our-health

cx http://www.ncbi.nlm.nih.gov/pmc/articles/PMC2518033/

cxi http://www.westonaprice.org/vitamins-and-minerals/salt-and-our-health

cxii http://www.westonaprice.org/vitamins-and-minerals/salt-and-our-health

cxiii http://www.westonaprice.org/vitamins-and-minerals/salt-and-our-health

cxiv http://www.scientificamerican.com/topic.cfm?id=heart-disease

cxv http://www.natural news

cxvi http://www.scientificamerican.com/topic.cfm?id=genetics

cxvii http://www.healingnaturallybybee.com/articles/salt7.php

cxviii How to Kill Parasites & Bacteria With Salt | eHow http://www.ehow.com/how_5661481_kill-parasites-bacteria-salt.html#ixzz2T6fHwczD

cxix http://www.dummies.com/how-to/content/understanding-the-transmission-of-nerve-impulses.html

cxx http://www.4bonehealth.org

cxxi http://science.jrank.org/pages/5948/Salt.html#ixzz2RCxvyw7z

cxxii http://science.jrank.org/pages/5948/Salt.html#ixzz2RCxvyw7z

cxxiii http://science.jrank.org/pages/5948/Salt.html

xxxv This section is from the book "The Hygienic System: Orthotrophy", by Herbert M. Shelton. Also available from Amazon:Orthotrophy.

cxxv http://www.answerbag.com/q_view/1570463#ixzz2QGqmZGHI

cxxvi http:www.buzzle.com/article/interesting-facts-about-sodium.html

cxxvii http//www.wikipedia.org

cxxviii http//www.wikipedia.org

cxxix http://www.westonaprice.org/vitamins-and-minerals/salt-and-our-health

cxxx http://www.prweb.com/releases/2011/11/prweb8948656.htm

cxxxi http://www.redorbit.com/news/health/1112428513/salt-consumption-debate-too-much-or-too-little/

cxxxii http://www.mayoclinic.com/health/low-blood-sodium/AN00621

cxxxiii www.hbci.com/~wenonah/hydro/al.htm

cxxxiv http://lpi.oregonstate.edu/infocenter/minerals/potassium/

cxxxv *https://www.khanacademy.org/science/biology/human-biology/v/sodium-potassium-pump*

cxxxvi *http://www.jonbarron.org/article/pillar-salt*

cxxxvii *http://www.tcolincampbell.org/courses-resources/article/salt-phobia/browse/1/?tx_ttnews%5BbackPid%5D=76&cHash=efd85ee330cd247adb2b2f54a 1fc5137*

cxxxviii *July 8, 2011 http://www.scientificamerican.com*

cxxxix *http://www.cnn.com/2011/HEALTH/05/03/salt.heart.attack/index.html*

xxxv *This section is from the book "The Hygienic System: Orthotrophy", by Herbert M. Shelton. Also available from Amazon:Orthotrophy.*

cxli

http://www.naturalnews.com/029053_salt_consumption_public_health.html#ixzz2OgC K4NfM

cxlii *Dr. David Brownstein, author of Salt Your Way to Health*

cxliii *http://www.naturalnews.com/031608_table_salt_sodium.html*

i1. Dr. M. Ted Morter, JR., M.A., "Your Health Your Choice."
2. Dr. Bernard Jensen, Ph.D., "The Chemistry of Man."
3. Harvard Health Publications, "Salt and your health, Part I: The sodium connection."
http://www.health.harvard.edu/newsletters/Harvard_Mens_Health_Watch/2...
4. Material Safety Data Sheet,. "Potassium Chloride."
http://www.jtbaker.com/msds/englishhtml/p5631.htm
5. International Programme on Chemical Safety,. "Potassium Chloride."
http://www.inchem.org/documents/pims/pharm/potasscl.htm

cxlv *http://www.naturalnews.com/table_salt.html*

cxlvi *http://www.naturalnews.com/031608_table_salt_sodium.html#ixzz2UA2hhuxx*

ii http://www.naturalnews
iii http://www.naturalnews

cxlix

http://www.naturalnews.com/038338_Himalayan_Salt_barter_items_survival.html#ixzz 2Og355QLU

cl *http://www.naturalnews.com/034528_salt_intake_diet_stroke.html#ixzz2OgApBUE3*

cli *http://www.naturalnews.com/034528_salt_intake_diet_stroke.html#ixzz2OgApBUE3*

clii *http://www.livestrong.com/article/176061-health-benefits-of-rock-salt/*

cliii *http://www.naturalnews.com/034528_salt_intake_diet_stroke.html#ixzz2OgApBUE3*

cliv *http://www.imi.com.sg/?page_id*

clv *http://anh-europe.org/news/not-worth-the-cash-report-says-unrefined-salts-as-dangerous-as-refined-salt*

clvi *http://www.globalhealingcenter.com/natural-health/how-safe-is-fluoride/*

clvii http://naturaldentistry.us/1378/the-dangers-of-fluoride/

clviii http://thyroid.about.com/od/symptomsrisks/a/Why-Are-So-Many-People-Getting-Thyroid-Disease.htm

clix http://www.cheeseslave.com/top-10-dangers-of-fluoride/

clx http://www.hoaxorfact.com/Health/health-warning-fluoride-toothpaste-dangerous.html

clxi Dr. Kim "Nine Secrets of Health"

clxii Dr. Kim "Nine Secrets of Health"

clxiii Dr. Kim "Nine Secrets of Health"

clxiv http://www.westonaprice.org/vitamins-and-minerals/salt-and-our-health

clxv http://www.westonaprice.org/vitamins-and-minerals/salt-and-our-health

clxvi http://www.westonaprice.org/vitamins-and-minerals/salt-and-our-health

clxvii http://www.westonaprice.org/vitamins-and-minerals/salt-and-our-health

clxviii http://www.westonaprice.org/vitamins-and-minerals/salt-and-our-health

clxix http://www.westonaprice.org/vitamins-and-minerals/salt-and-our-health

clxx http://www.westonaprice.org/vitamins-and-minerals/salt-and-our-health

clxxi http://www.westonaprice.org/vitamins-and-minerals/salt-and-our-health

clxxii http://www.westonaprice.org/vitamins-and-minerals/salt-and-our-health

clxxiii http://www.westonaprice.org/vitamins-and-minerals/salt-and-our-health

clxxiv http//www.wikipedia.org

clxxv Forensic Sci Int. 2008 Jan 30;174(2-3):173-7. Epub 2007 May 3 clxxv.

clxxvi http://www.clinchem.org/content/50/11/1961.full

clxxvii http://www.riordanclinic.org/laboratory/about/

clxxviii http://ajcn.nutrition.org/content/43/3/438.full.pdf

clxxix http://www.plosone.org/article/info%3Adoi%2F10.1371%2Fjournal.pone.0028824

clxxx http://www.plosone.org/article/info%3Adoi%2F10.1371%2Fjournal.pone.0028824

clxxxi http://www.plosone.org/article/info%3Adoi%2F10.1371%2Fjournal.pone.0028824

clxxxii http://www.ncbi.nlm.nih.gov/pubmed/11694882

clxxxiii http//www.wikipedia.org

clxxxiv http//www.wikipedia.org

clxxxv http://www.plosone.org/article/info%3Adoi%2F10.1371%2Fjournal.pone.0028824

clxxxvi http://www.ehow.com/way_5534258_alcohol-detox-sweats.html

clxxxvii http://www.medicinenet.com/night_sweats/page5.htm

clxxxviii http://www.medicinenet.com/night_sweats/page4.htm

clxxxix http://www.plosone.org/article/info%3Adoi%2F10.1371%2Fjournal.pone.0028824

cxc http://www.medicinenet.com/night_sweats/page2.htm#cancer

cxci http://www.medicinenet.com/night_sweats/page2.htm#cancer

cxcii http://www.medicinenet.com/night_sweats/page5.htm#hormone_disorders

cxciii http://www.medicinenet.com/night_sweats/page5.htm#hormone_disorders

cxciv http://www.medicinenet.com/night_sweats/page2.htm

cxcv http://www.plosone.org/article/info%3Adoi%2F10.1371%2Fjournal.pone.0028824

cxcvi http://www.plosone.org/article/info%3Adoi%2F10.1371%2Fjournal.pone.0028824

cxcvii http://en.wikipedia.org/wiki/Water_retention_(medicine)#cite_note-7

cxcviii http://www.webmd.com/allergies/medications-that-may-cause-swelling

cxcix http://www.madehow.com/Volume-2/Salt.html#b#ixzz2OgIs0W8m

cc http://www.westonaprice.org/vitamins-and-minerals/salt-and-our-health

cci https://en.wikipedia.org/wiki/In_viv

ccii http://www.sebiology.org/publications/Bulletin/July05/salinity.html

cciii http//www.wikipedia.org

cciv http//www.wikipedia.org

ccv [source: Salt Institute].

ccvi http://www.dailymail.co.uk/news/article-2312547/Knewra-salt-Inside-large-mosque-electric-railway.html#ixzz2RU883kb0

ccvii http://geography.about.com/od/physicalgeography/a/oceancurrents.htm

ccviii http://geography.about.com/od/physicalgeography/a/oceancurrents.htm

ccix http://www.esajournals.org/doi/abs/10.1890/110004

ccx http://geography.about.com/od/physicalgeography/a/oceancurrents.htm

ccxi http://science.nasa.gov/earth-science/oceanography/physical-ocean/salinity/

ccxii https://sites.google.com/site/healthpartisan/salt-alert---aluminium

ccxiii www.livestrong.org

ccxiv www.wickipedia.org

ccxv
http://www.naturalnews.com/038617_table_salt_coarse_kosher.html#ixzz2Og7XNRdp

ccxvi http://www.newton.dep.anl.gov/askasci/mole00/mole00093.htm

ccxvii http://annieshealthtalk.hubpages.com/hub/How-to-Properly-Clean-an-Open-Wound

ccxviii How It Works: Water Softener - Popular Mechanics

ccxix http//www.wikipedia.org

ccxx http://www.saltinstitute.org

ccxxi mad-science.wonderhowto.com

ccxxii http://news.nationalgeographic.com/news/2007/09/070913-burning-water_2.html

www.ingramcontent.com/pod-product-compliance
Lightning Source LLC
Chambersburg PA
CBHW051706170526
45167CB00002B/562